IMPACT OF INDUSTRIAL EFFLUENT WATER ON AGRICULTURAL LAND

Impact of Industrial Effluent Water on Agricultural Land

DR. S. JAGADEES PANDI

ABHIJEET PUBLICATIONS
DELHI 110094

ABHIJEET PUBLICATIONS
2/46 Tukhmeerpur Extension
Delhi-110 094
Phone: 011-22960492 (Fax), 65698474
E-mail: abhijeetpublication@gmail.com

IMPACT OF INDUSTRIAL EFFLUENT WATER ON AGRICULTURAL LAND

First Published 2008

ISBN 978-81-89886-75-2

PRINTED IN INDIA

Published by J.K. Singh for Abhijeet Publications, Delhi-110 094, Lasertypeset by Gaurav Graphics, New Delhi and Printed at Himanshu Printers, Delhi.

Preface

The contamination and quality of irrigation water is of the main concern especially in the regions with limited water resources. In such region not only, the water resources should wisely be utilized at the same time should be prevented from contamination. The Effluent water discharged by all kind of industries in the most important sources of soil and water contamination. The supply of water effluent through rivers has some serious repercussions on health effects as well as affects land productivity. Crop production suffers from using contaminated irrigation water form both surface sources and from ground water aquifers. In this context an attempt is made to study the socio-economic impact of effluents discharged by the industries in Coimbatore Region, Tamilnadu.

I am greatly indebted to the Board of Management of Sree Saraswathi Thyagaraja College, Pollachi, whose constant and affectionate interest in my progress has been the source of inspiration for this publication. I am happy to extend my profound thanks to Dr. V. Kulandaiswamy, Principal, Sree Saraswathi Thyagaraja

College, Pollachi, for his inspiring Guidance and the interest he evinced in my research work.

I extend to my sincere thanks to all the staff members of Commerce Department, Sree Saraswathi Thyagaraja College, Pollachi, for their worthful guidance. I Place on record my deepest appreciation to my family members whose affectionate interest and continued support to all through my research activities.

S. JAGADEES PANDI

Contents

List of Abbreviations

ANOVA - Analysis of Variance
BOD - Biological Oxygen Demand
CETP - Common Effluent Treatment Plant
COD - Chemical Oxygen Demand
CPCB - Central Pollution Control Board
EC - Electrical Conductivity
FAO - Food and Agricultural Organisation
IETP - Individual Effluent Treatment Plant
LEC -- Loss of Ecology Authority
MEC - Marginal External Cost
MPC - Marginal Private Cost
MPB - Marginal Private Benefit
MSB - Marginal Social Benefit
MSC - Marginal Social Cost
NGO - Non Governmental Organisation
PAC - Public Accounts Committee

PWD	-	Public Works Department
SSI	-	Small Scale Industries
TDS	-	Total Dissolved Solids
TSS	-	Total Suspended Solids
TNPCB	-	Tamilnadu Pollution Control Board
UNDP	-	United Nations Development Programme
UNEP	-	United Nations Environment Programme
UNIDO	-	United Nations Development Organization

1

Introduction and Design of Study

The economy and the environment are not separate entities. Economy forces changes on the environment, which in turn react back forcing unforeseen changes in the economy. The environment can be defined as the physical surrounding of man of which he is a part and on which he depends the activities like physiological functioning, production and consumption, where as the economy focuses on the well-being of an individual and societies. The physical environment of human stretches from air, water and land to natural resource like energy carriers, soil and plants. The relationship between environment and economy is multi-fold and a multi faced with a qualitative well as a quantitative aspect to it. For example, the definition of pollution in economics is based not only on the physical effect on the environment but crucially on the human response to the physical effect. The physical effect can be biological (or) chemical while the human response is a result of loss of welfare[1]. Thus physical scientists base their perception of pollution solely on the physical effect on the environment, while economists rely on perceived utility losses to recognize pollution.

Industrial pollution has been a major factor causing the degradation of the environment around us, affecting the water we use, the air we breathe and the soil we live on. This Industrial environmental pollution is the major problem associated with rapid industrialization along with urbanization and raises in standard of living of the people. For developing country, industrialization is must and still this activity very much demands to build self-reliant and in uplifting nation's economy. Therefore rapid pace of industrialization not only increases the output level but also causes a serious threat to human welfare. Water is polluted not only by industries but also by households. Both industries and household's waste water contain chemical and biological matter that imposes high demands on the oxygen present in water. The polluted water contains low levels of dissolved oxygen as a result of the heavy Biological Oxygen Demand (BOD) and Chemical Oxygen Demand (COD) placed by industrial and household waste materials discharged into water bodies and water systems, both above and below the earth surface. In addition to levels of dissolved oxygen in water, industrial wastes which contains chemicals and metals, that are directly harmful to human health and the eco system[2].

Environmental Degradation

In the literature of *Wikipedia*[3] and Encyclopedia environmental degradation is the deterioration of the environment through depletion of resources such as air, water and soil. When the environment becomes less valuable (or) damaged, environmental degradation is said to occur. There are many form of environmental degradation. When habitats are destroyed biodiversity is lost (i.e.) national resources are depleted, the environment is hurt. They can occur naturally (or) through human processes. Development activities are measured in terms of natural products, which in turn are defined as production of goods and services during accounting period.

However, certain environmental functions, which are crucial for economic performance and generation of human welfare such as provision of natural resources to production and consumption activities, waste absorption by environmental media and environmental services of life support and other human amenities, are taken into account only partly in conventional accounts. The scarcities of natural resources threaten the sustained productivity of the economy and economic production and consumption activities. These activities impair environmental quality by overloading natural environ with wastes and pollution.

The environmental consequences of development tend to offset many benefits that may be accruing to individuals and societies on account of rising incomes. There are direct costs on the health of individuals, their longevity and on quality of life on account of deterioration in environmental quality to mention a few more importantly, the environmental damage can also undermine future attainments and productivity, if the factors of production are adversely affected. Therefore the private and social costs of the use of the natural resources and the degradation of the environment may be taken into for the sustainable development in the conventional accounts[4].

The key environmental challenges that the country faces relate to the nexus of environmental degradation with poverty in its many dimensions, and economic growth. These challenges are intrinsically connected with the state of environmental resources such as land, water, air and their flora and fauna. The proximate drivers of environmental degradation are population growth in appropriate technology, consumption choices and poverty leading to changes in relation between people and its system and development activities such as intensive agriculture, polluting industry and unplanned urbanization. In addition to the above matters, Institutional failures referring to unclear (or) insufficiently enforced rights of access to and use of environmental resources result in environmental degradation because third parties primary experience inputs of such degradation, without cost to the persons responsible for the damage.

Environmental, degradation is a major casual factor in enhancing and perpetuating poverty, particularly among in rural poor, when such degradation impacts soil fertility, quality of water, air, forests resources, wild life and fisheries. The poor are always more vulnerable of loss of resilience of eco system. Economic growth has a dichotomous relationship with environmental degradation, on the one hand growth may result in environmental degradation through use of natural resources. On the other hand economic growth permits improvement in environmental quality by making available the necessary resources for environmental investments for improved environmental behaviour, institutional and policy changes.[5]

Land Degradation

The term Land refers to all natural resources which contribute to agricultural production including livestock production, and afforestation. It covers climate and water sources, land form, soil and vegetation, including both grass land resources and forests. Land is the basic natural resource that provides habitual and sustenance for living organisms, as well as being major forces of economic activities.

Land degradation defines as a loss of the productive capacity of land to sustain life, (or) degradation of land refers to loss of its potential production capability. It is the temporary (or) permanent lowering of the productive capacity of land. The Vegetation and Management Act, India (1995) states that the phrase includes soil erosion, rising water tables, the expression of salinity, mass movement of gravity of soil (or) rock, stream bank in stability and a process that results in a declining water quality.

As a result of degradation of soil quality and also its loss for effective use, and become the most serious threat to agricultural productivity and bio diversity. The topsoil degrades due to natural process and human activities. The functional capabilities of soil deteriorate from activities related to agriculture, forestry and industry. On the other hand, urban sprawling and infrastructure development cause loss of available land. Natural calamities such as cyclones and floods cause land loss and can also deteriorate functional capabilities of soil. Soil degradation in the coastal area results from unplanned land use, intrusion of saline water. So the land degradation problem is based on multi-sectored or multilayered.

Land degradation can be considered in terms of the loss of actual or potential productivity (or) utility as a result of natural (or) anthropic factors; it is the decline in land quality (or) reduction in its productivity in the context of productivity. Land degradation results form a mismatch between land quality and land use among physical processes are a decline in soil structure leading to crusting, compaction, erosion, desertification, environmental pollution and unsustainable use of natural resources[6]. Significant chemical processes include acidification, leaching salinization and fertility depletion. Biological processes include reduction in totally and bio-mass carbon and decline in land biodiversity. The latter comprises important concerns related to eutrophication of surface water, contamination of ground water and emission of trace gases from terrestrial/aquatic ecosystem to the atmosphere. Soil structure is the important property that affects all these degradation processes. Thus land degradation is a bio-physical process driven by socio-economic and political causes.

The findings of the pioneering study[7] was sponsored by three united nations agencies (FAO, UNDP and UNEP) estimated the severity and costs of land degradation in South Asian Countries such as India, Pakistan, Bangladesh, Afghanistan, Nepal, Sri Lanka and Bhutan. The study disclosed that these countries are losing at least US $10 billion annually as a result of losses resulting from land degradation. The assessment of the economic costs of land degradation estimate that at $ 9.8 to 11 billion US dollar in a year, of which due to water erosion $5.4 billion wind erosion $ 1.8 billion, fertility decline $ 0.6 billion, water logging $ 0.5 billion and salinisation $ 1.5 billion. A rough estimate of soil erosion and sedimentation for India reveals that about 5300 million tonnes of top soil are eroded annually and 24 percent of this quantity is carried by rivers as sediments and deposited in the sea and nearly 10 percent is deposited in reservoirs reducing their storage capacity by 2 percent. As for as the water logging is concerned the canal command area constitutes 48 percent of the total water logged area, and 4.5 percent of the total salt affected area in India. Few states like Andhra Pradesh, Tamilnadu, Orissa, Punjab and Gujarat, canal irrigated area occupies 100 percent of the total water logged area[8].

A study by Sehgal and Abro (1994) estimated that around 162 million hectare of soil is degraded by soil erosion, includes wind and water erosion, the saline and alkaline soil accounts for 10.1 million hectare, degradation due to water logging 11.6 million hectare, shifting cultivation accounted for 9 million hectare of land[9].

There are many driving forces compete the society to over exploit natural resources (or) degradation for land. The main causes are deforestation, ignorance of soil conservation practices, extension of cultivation to marginal lands, improper crop rotation imbalanced fertilizer use, rapid growth of industry, paucity of land and economic pressures with poverty[10]. In this section an attempt is made to analyze the causes and effects of land degradation, as such the above said factors are grouped as viz. forest resources and water resources.

The major causes of deforestation are discussed below:

i) Over-exploitation by communities residing in the vicinity of forests for subsistence (or) economic use such as collection of fuel wood, fodder, non-timber, forest produce etc.

ii) Over-grazing by livestock.

iii) Commercial extraction of timber and non-timber forest produces by state agencies.

iv) Unregulated urbanization and increased consumption needs of urban population.

v) Rapid Industrialization

vi) Diversion of forest land to no-forest use such as mining industry, timber based industry, paper and pulp industry and so on.

vii) Submergence of forest land in river valley projects.

viii) Conversion of forest land for agriculture purposes including encroachments.

The increasing factor of the deforestation is serious threat of Bio–diversity. The loss of bio-diversity increases marginalized and impoverished many tribal and other forest dependant communities, which directly depend on various components of bio-diversity for food, drink, medicine, culture, spirituality etc. Even among the poor women are affected more than men. In many societies, women bear the major responsibility of collection of fuel, fodder, water and such basic subsistence needs. As these resources are depleted, drudgery of women increases and to some extend they have to walk for more distances to collect these resources. This causes serious health problems for women. The indirect impact of degradation such as severe drought (or) deforestation–induced floods also hits the poor hardest.

The major causes of degradation of land from water sources are below:

The protection and management of ground water which is one of the most valuable natural resources are depleted because of contamination by wastes. Once the ground water is contaminated and it is difficult to restore to its original quality. The economic survey of

India (1995) lamented that 95 percent of the water in India's 241 class II towns were polluted[11].

The quality changes of water resources because of the following factors:

1) Deforestation and poor land use practices in the catchments area, which disturb topsoil and vegetation cover resulting in increased infiltration rates, increased runoff, sediment transport and deposition in rivers and storage reservoirs.

2) Over abstraction of surface water sources at the upstream reduces the minimum flow required in the downstream sections for the sustenance of ecosystems and mangroves.

3) Excessive water withdrawal for various seasons such as irrigation, industry and domestic consumption. It induces saline water intrusion into fresh water aquifer resulting changes on ground water quality with increased TDS.

4) Agricultural drainage, which is carrying residues of chemical fertilizers and pesticides affects the water quality promoted weed growth, and renders the water resources unfit for other uses.

5) Encroachment of agricultural land and watersheds for urbanization and industrial development has impact on wetlands and important watershed areas and affects recharging areas and reservoir capacities.

6) The natural drainage and water flow of many important rivers has been altered due to the withdrawal of water and construction of big dams.

7) Water pollution due to discharge of untreated/

partially treated industrial and municipal waste water into water sources deplete dissolved oxygen and affects fish and other aquatic life.

Various Process of Land Degradation

The various processes of land degradation are grouped into six classes:

a) Water erosion covers all forms of erosion by water; including sheet and rill erosion and gullying. Human induced intensification of land sliding caused by vegetation, clearance road construction etc., is also included.

b) Wind erosion refers to loss of soil by wind, occurring primarily in dry regions.

c) Soil fertility decline refer to what is more precisely described as deterioration in soil, physical, chemical and biological properties whilst decline in fertility is indeed a major effect for erosion. The main process involved are; lowering of soil organic matter, degradation of soil physical properties, adverse change in soil nutrient resources building up toxicities, primarily acidification through incorrect fertilizer one.

d) Water logging is the lowering in land productivity through the rise in ground water close to the soil surface; water logging is linked with salinization both being brought about by incorrect irrigation management.

e) Salinization refers to all of soil degradation brought about by the increase of salts and in the soil. It covers both salinization and alkalization; as human induced processes, these occur mainly through incorrect planning and management of irrigation schemes.

f) Lowering of the water table is a self explanatory form of land degradation, brought about through tube well pumping of ground water for irrigation exceeding the natural recharge capacity

In India the intensive farming practices after the green revolution, particularly wheat and rice have virtually mixed nutrition from the soil. Due to heavy use of fertilizers, excess nitrates have reached into ground water and contamination of ground water with nitrates has increased dramatically. As such, the cultivable lands become suck by over application of chemicals. Similarly intensive agriculture causes to improper use and maintenance of canal irrigation has contributed significantly to the soil degradation problem. Extension of canal irrigation to avoid semi-arid areas has resulted in water logging and salination. According to the National Remote Sensing Agency and Forest Survey of India 60 million of the total area under cultivation is substantially degraded most of the damage was in the form of loss of topsoil. Large scale ecological losses were reports in Crop Land, Grass land and Forest Land such as soil erosion, soil alkalinity and salinity, nutrient deficiency, water logging and contamination of ground water. Irrigation is considered as the principle means of water loss from the natural system and it leads to ground water depletion[12].

Table 1.1: Soil Degradation in India

Area in Million Hectures

S.No.	*Type*	*1980*	*1985*	*2005*
1	Soil Erosion	150.00	141.2	103.16
2	Saline & Alkaline Soil	8.0	9.2	5.94

3	Water logging	6.0	8.5	14.30
4	Shifting Cultivation	4.4	4.9	Nil
5	Other reasons	Nil	Nil	23.41
Total Degradation of Land		168.4	175.1	146.82

Source: Reports from Ministry of Agriculture and Co-operation Govt. of India.

Effects of Land Degradation

The various effects caused by degradation of land are discussed below.

Total Abandonment of Land

Where degradation reaches the severe degree, formerly productive land must be abandoned. *i)* Salinization patches e.g. In India and Pakistan where the outlines of fields can be still seen in areas where whole farming have been abandoned. Large tracts of land have totally gone out of production. *ii)* Total removal of the soil from steep slopes, e.g. in parts of the hill country of Sri Lanka, where farmer takes cultivation has been abandoned and the land colonized by grass land of low productive potential[13].

Reduced Crop Yields

Where land has been subject to light (or) moderate degradation the same level of inputs will give longer outputs. A study in Andra Pradesh, based on artificial removal of topsoil showed clear and strong relations between topsoil depth and crop yields. The slope of the loss curve being greater for high rainfall than low rainfall years[14]. Reduction in crop yield due to salinity has been well researched, with percentage of losses differing from as between salt sensitive and tolerant crops. For range lands, a reduction in livestock productivity to 10-50 percent of its potential is estimated for decertified range lands of Pakistan. Forest productivity is also reduced on degraded land, although this is mitigated by the capacity of some tree spices to tolerate poor soils, and the recuperative effects upon soil of appropriate reforestation[15].

Increased Inputs and Greater Costs

The level of poverty of many farmers is such that they cannot accept the consequence of reduced yields. Instead, they must attempt to maintain their food supplies from the degraded land, by means of increased inputs. This is now widely the case where attempts are made to combat soil fertility decline by increased inputs of fertilizers. Another response is to attempt to maintain livestock numbers despite a reduced carrying capacity of pastures, thus leading to a vicious circle of further degradation.

Reduced Responses to Inputs

It is now accepted that fertilizers are best utilized by application of low to moderate amounts, whilst seeking to obtain high responses. Land degradation, particularly the lowering of soil organic matter, has the opposite effect, that of lowering fertilizer responses.

Reduced Productivity on Irrigated Land

A specific case of lower crop yields and reduced responses to inputs occur on the irrigated lands which are widespread in the region. These irrigation systems have been established at high cost, whether of capital, as in tube wells and the large reservoir and canal schemes, or labour, as in the cases of hand dug wells and earth dams. Lowered productivity, as a result of soil fertility decline, water logging and salinization, reduces the benefits from irrigation, leading to less efficient use of capital and labour inputs.

Loss of Flexibility in Land Management

Reduced crop yields can force farmers to grow only basic food crops, particularly cereals. Again there is a feedback effect, since continuous cereal production causes decline in soil fertility.

Greater Risk

Degraded land is less resilient, less able to recover from recurrent disasters, such as drought. One of the major effects of erosion is reduced water-holding capacity of the soil. Increase in risk places constraints on land management, making farmers reluctant to use up scarce capital on fertilizers.

Loss of Water for Irrigation

An off-site effect of deforestation and erosion of watershed areas is destabilization of river flow regimes, causing flooding after rains and reduced flows in subsequent dry periods. Where there are downstream irrigation systems, this reduces water availability at times when it is most needed.

Lowering of the Water Table Increases Irrigation costs, and can make groundwater totally unavailable to farmers with small landholding, who cannot afford deep tube wells.

Diversion of Resources to Reclamation

As already noted with respect to fertilizer inputs, pressure of need forces farmers to make great efforts in an attempt to maintain production. Thus they may construct gabions across gullies, or build terraces with fines of stores, requiring large amounts of labour, or construct deeper tube wells. All such inputs, whether of labour or capital, carry an opportunity cost, that of the alternative, productive, uses to which the resources could have been applied[16].

Review of Related Studies

Mukherjee and Nelliyat[17] (2006) found that continuous disposal of industrial effluents on land, which has limited capacity to assimilate the pollution load, has lead to ground water pollution. A case study is Mettupalayam Taluk of Tamilnadu identified that ground water quality of shallow open wells surrounding the industrial locations has deteriorated, and the application of polluted ground water for irrigation has resulted in increased salt contents of soil. Since the farmers in this study area had already shifted their cropping pattern to salt tolerant crops viz. Jasmine, Tobacco, etc., and substituted their irrigation source form shallow open wells to deep bore-wells and or river water. More over the study observed that with the rise in concentration of electrical conductivity of ground water samples, revenue from banana cultivation has gone down.

Ratna Reddy and Bhagirath Behara[18] (2006) assessed those economic costs of Industrial water pollution in the rural communities in terms of losses to agricultural communities human health, and livestock. They conducted a study on two villages, one was a pollution affected and the other was not affected, located in one of the Industrial estate in Andhra Pradesh. The cost estimates revealed that the impact of Industrial pollution on rural communities is quite substantial in monetary terms. The paper argued that the compensation principle might work if the estimate of damage is realistic. Further passing of laws and creating institutional structures are necessary but not sufficient to address the environmental problems. Policies should be implemented in their right perspective and institutions should be strong enough with more autonomy and powers.

Srinivas[19] (2006) reported that the effects of untreated effluents of small and medium scale Industries discharged into the river Khari in Ahmedabad district. This has resulted into contamination of ground water up to the depth of 183 meter. About 110 villages of downstream areas are drinking polluted color water for 20 years. The water pollution has generated a variety of social problems for the people. The study disclosed that for more than two decades the people have been suffering from health-related problems which include skin diseases, stomach and internal ailments and bronchial problems. Deteriorating soil fertility has had an impact on agricultural produce, which in turn reduced income of the farmers. Moreover livestock casualties and reduction in milk yield have adversely affected the village economy.

A study[20] (2004) was undertaken in Kanpur and Varanasi of Uttar Pradesh to assess the untreated waste water discharged by the sewage treatment plants. The study suggested a definite adverse impact on the environmental quality of the disposal area. The critical level of the heavy metals in the soil for agricultural crops are found to be much higher than those observed in the study areas receiving no effluents The sludges had Cadmium, Chromium, and Nickel level were above tolerable levels as prescribed for agricultural and land application.

A study was carried out in Malawi[21] (2003) to assess the extent of pollution in a receiving river as affected by industrial effluents. The study shows that the effluents from the industries had a big impact on the water quality on the receiving river. The result of the study shows, effluents were very acidic in both dry and rainy season. The water, after the effluents receiving points was acidic and the level of the other parameters was high and especially downstream. The dissolved oxygen, alkalinity, suspended solids, and chloride were relatively high. The study suggested that the effluents were being discharged into the river have considerable negative effect on the quality of the water in the river and as such, the water was not good for human use and agricultural work. The study suggested there is need to install treatment plant for all industrial works.

Chellappan Sekar[22] (2003) study in Tamil Nadu, where the pollution of ground water is high due to Industrial effluent shows that the water resources have created some external effects on land quality, land value, crop yield, loss of man power, employment etc. The externalities caused by the changes in the qualities of water resources due to mismanagement and misuse included the common health disorders. Farmers had incurred a total amount of Rs. 518 per annum household towards the cost of physician and medicine and loss in time and management.

Reduction in animal population, poor health status reduction in milk yield were and the external effects on animal due to degradation of ground water, land selling, diversion of farm lands for non-agricultural activities, labour movements towards non-farm sector and poor economic status were the consequences of pollution externalities observed.

Nataraj *et al*[23] (2003) conducted a study in Vrisabavati river basin in Bangalore, identified that there was a poor quality of ground water and lower yield of crops The productivity of rice was lower by 62 percent and sugarcane yield lower by 47 percent in the affected farms compared to normal farms. The field irrigated with polluted water is loaded with heavy metals like ferrous, and Manganese causing hardening the soil upon irrigating polluted water to the vegetables, oilseeds, maize and pulses. The symptoms of wilting and finally crop losses were observed as these crops are salts sensitive.

Malik and Negi[24] (2003) studied the impact of distillery effluent on water quality and phytoplankton on the river saving revealed that distillery effluent contains a large amount of dissolved organic matter and causes a severe damage to a actuate life. The abundance of phytoplankton has been affected by the highly polluted condition of river as indicated by decreasing trends in phytoplankton in the study area.

Jothimani and Baskaran[25] (2002) carried out a study to elucidate an appropriate dilution of dying factory effluent for irrigation on agricultural crops and to assess the changes in soil, electrical conductivity and organic carbon during dyeing factory effluent on irrigation. The results of the study showed that the effluent could be safely used for irrigation at proper dilutions in combination with NPK.

Alam[26] (2003) studied the Physio-Chemical and toxicology of Industrial effluent around Delhi by collecting some water samples and found that the water quality from the parameters from the localities located on the side of Yamuna River was very poor.

Udaya Suriya *et al*[27] (2003) conducted a study to assess the impact of continuous effluent irrigation on soil and ground water by paper industry on the Cauvery river basin. The study had found that the paper mill effluent had increased the soil pH, electrical conductivity in irrigated soils. The potential salinity of the ground water in effluent irrigated field was increased three terms over control and the values crossed the safer limits in many places indicating that the ground water was polluted with chlorides and sulphates.

Patel and Pandey[28] (2003) studied the sustainability of industrial effluents for irrigation in an industrial zone of Gujarat revealed that industrial effluent collected from different cities were found contaminated with all major polluting element. It contained high level TSS, COD, BOD and the higher standards permissible limit for irrigation. The well water also had shown salinity and alkalinity hazardous.

Behara and Ratna Reddy[29] (2002) studied the environment and accountability and the impact of Industrial pollution on rural communities in Medak district of Andhra Pradesh, that was none of the oldest most environmentally degraded area. The study had found that the entire village was suffering from various diseases arising out of water pollution. From this study, it was observed that most of the diseases of water-borne, such as skin infections, defective vision, fever, teeth corrosion, joint pains, and loss of appetite, respiratory diseases and diarrhoea.

Fazeli *et al*[30](1998) analyzed effects of the irrigation by the effluents discharged by the nearest paper mills in Nanjungud of Mysore District, Karnataka. The study found that the accumulation of heavy metals in the soil and different path of the paddy crops. The study had found that the chemical and biological oxygen demands of the above tolerance limits set by Indian Standards. There was a greatest accumulation of metals in the leaves and roots of crops. The study suggested that, since in many tropical countries the common diet of people is rice, the accumulation of toxic heavy metals in rice may lead to health disorders.

Rao *et al*[31](1993) conducted a study in Bandi River for assessing the impact of textile industrial effluents on agricultural land. The investigating study found that the effluents were highly saline, coloured and sodic in nature resulting in the river and well water unsuitable for irrigation. The biological activity of the discharge of effluents of textile mills posed a serious threat to the health of these helpless farmers and their crops. To investigate the effect of land–use on water quality, selected physio-chemical parameters, temperature, velocity, total alkalinity, total phosphate, chloride was analyzed during monsoon and post monsoon periods.

Yadav[32] (1990) observed that the need for treated effluent water for agricultural land explained that there is an urgent need to utilize the alternate sources of water for agriculture in order to the demand. According to him increasing industrialization and urbanization is its wake of the major problem of safe disposal of waste water. Good quality of water is inadequate when for normal living and is getting polluted due to industrial discharge including those of paper, textile, tannery, distilleries and other industries. The study suggested that efforts should be made in future for proper treatment and safe disposal of industrial effluents and use in agricultural land in order to increase our food grain production and enhancing environment quality.

Aparna[33] (1989) conducted a study on water pollution by tanneries in Dindigul District reported that the level of various Physio-Chemical factors indicative of pollution were found to exceed the quality standards and affects HO_2 quality pattern. The ground water was found in the low quality near the effluent lagoon and level of pollutants was found to decrease with increasing distance of the water source from the effluent lagoon.

The Present Study

The contaminated water for irrigation is the main concern especially in the regions with limited water resources. In such region not only, the water resources, it should wisely be utilized at the same time should be prevented from contamination. The Effluent water discharged by all kind of industries in the most important sources of soil and water contamination. The Effluent water usually contains high level of hazardous material, which removal of them may not be possible with routine treatment.

The supply of effluent water through rivers has some repercussions on health effects as well as affects land productivity. Crop production suffers from using contaminated irrigation water form both surface sources and from ground water acquifers. From the available studies and reports, the effluents discharged into the Noyyal river by the Small Scale Textile Units in

Tiruppur in Coimbatore District affects the soil, surface, and ground water for both drinking and irrigation purposes. This in turn creates negative externalities viz., damages the agricultural productivity, drinking water scarcity and causes to increase in social cost of the farmers and the rural folk.

In this regard it is indispensable for a research study to be carried out in the affected regions by the Noyyal River in Tamilnadu.

Objectives and Methodology

The specific objectives of the study are; *i)* to understand the nature and impact of the effluent water by the Noyyal river; *ii)* to review the socio economic status of the farmers affected by the effluent water; and *iii)* to provide suggestive measures towards improving the existing environmental condition.

The farmers, who had been affected by the Noyyal river, from Tiruppur Taluk in Coimbatore District and Perundurai Taluk in Erode district of Tamilnadu. With regard to sampling cluster sampling technique is adopted. In the first stage the study area is divided into two zones. Each zone comprises five revenue villages. The farmers who have been cultivating the land with in the radius of 2km from the Noyyal River were the sample farmers. In the second stage 75 farmers from each Zone were chosen. A total of 150 farmers were taken for the study.

The study is mainly designed to use primary data. A structured-cum-schedule was canvassed to the selected respondents. In addition in this, secondary data was collected from the respective taluk offices, assistant statistical offices. Besides focal group discussion were made from the presidents of farmers association and local NGO's. The Collected data were fitted into the suitable parametric and non-parametric test such as ANOVA, MANN WHITNEY 'U' test, and Chi-squire test.

Variables of Study

The following variable to be used the study:

- ❖ The demographic variables of the study include age, number of children, family members like, father and mother types of family of the farmers;
- ❖ Details of agriculture viz., acres of land, different holdings of land, types of land possession, sources of irrigation, crops of cultivation, shifting pattern of crops, seasons of crops cultivation;
- ❖ Damage assessment of agriculture includes variation in yielding of crops, changes in quality of lands in physical terms, productivity of land etc.,
- ❖ Social cost of agriculture includes changes in the income from agricultural farming, diversification of agricultural farming, cost of agricultural inputs, marketing activities, loss of agricultural production, sale of land etc.,

Conceptual Framework

The following Concepts have been discussed in the present study.

Biological Oxygen Demand (BOD)

It is expressed as weight of oxygen consumed per

unit volume of water during a defined period of time at a defined temperature.

Common Effluent Treatment Plant (CETP)

In a Common Effluent Treatment Plant, the effluents from the different industries are treated using one universal treatment system. Common Effluent Treatment plants eliminate duplicity of treatment systems among the industries on the industrial estate and hence results in a reduction in the total capital required for construction of the industrial estate

Electrical Conductivity (EC)

It is the measure of the ability of an aqueous solution to convey an electric current. This ability depends upon the presence of iron, their total concentration, mobility and temperature.

Reverse Osmosis (RO) System

The main function of RO is to recovery of pure water from the effluent for re-use in textile wet processing operations.

pH

It is defined as the logarithms of the activity of hydrogen iron with negative sign.

Salinity

The amount of salt (Especially sodium chloride) in water

Total Suspended Solids (TSS)

Total Suspended Solids are the portion of solids that usually remains on the filter paper. They consist of Silt,

Clay, fine particles of organic and in organic matters, which is regarded as a type of pollution, because water high in concentration of suspended solid may adversely affect flora and fauna.

Total Dissolved Solids (TDS)

It is the measure of total inorganic salts and other substances that are dissolved in water.

Zero Discharge System

It is meant to treat all the incoming effluents, and leave nothing behind. The main aim is to recover usable material such as water and salt from the effluent and minimize the generation of waste. So that it can be safely stored on-site and can prevent the discharge from spoiling the environment.

REFERENCES

1. Senaca, J.J., Taussig, M.K (1984) *Environmental Economics*, Prentice-Hall Inc., New Jersey.
2. Govindarajulu (2003) "Industrial Effluent and Health status- A case study of Noyyal River Basin", *Proceedings of the Third International Conference on Environment and Health*, Chennai, December 15-17, pp. 150-157.
3. *Wikipedia* of Enclocycopedia available at www.wikepedia.org
4. Government of India (2006) "Compendium of Environment Statistics in India", Ministry of Statistics and Implementation Programme, pp. 1-3.
5. *Ibid.*
6. Bienroth, F.H., Eswaran, H., Reich, P.F. & Van Den Berg, E. (1994) "Land Degradation stresses in agrosystems" in (ed) S.M. Virmani, Katyal, Eswaran, H., & Abrol I.P., *Stressed Ecosystem and Sustainable Agriculture*, Oxford and IBH Publications, New Delhi.
7. Food and Agricultural Organisation–FAO(1976) "A Framework for Land Evaluation", *FAO Soils Bulletin*, Rome pp. 32.

8. Food and Agricultural Organisation of the United Nations (1994) "Land degradation in South Asia: Its severity, causes and effects upon people", World Soil Resources Report, Rome.
9. Sehgal, J. & Abrol, I.P. (1994) *Soil Degradation in India: Status and Impact,* Oxford and IBH Publishing Co., New Delhi.
10. Government of India, Ministry of Environment and Forest available at www.enfor.nic.in
11. Government of India (1995), Ministry of Finance, Economic Survey Report.
12. Ninan, K.N. (2002) "Watershed Development Programmes in India—A Review", Paper presented at the ISCO conference, Beijing.
13. *Op.cit.,* FAO (1994).
14. Vittal, K.P.R., Vijayalakshmi, K. and Rao, U.M.B. 1990. The effect of cumulative erosion and rainfall on sorghum, pearl millet and castor bean yields under dry farming conditions in Andra Pradesh, India. *Experimental Agriculture,* Vol. 26, pp. 429-39.
15. Ahmad, M. and Kutcher, G.P. (1992) "Irrigation planning with environmental considerations: a case study of Pakistan's Indus basin." World Bank Technical Paper 166. World Bank, Washington DC, 196 pp.
16. Joshi and Jha (1991) "An Economic Enquiry into the Impact of Soil Alkalinity and Water logging", *Indian Journal of Agricultural Economics,* Vol. 47, pp. 195-204.
17. Sachidhanantha Mukerjee & Prakash Nelliyat (2007) "ground water pollution and emerging environmental challenges of Industrial effluents irrigation—A case study in Mettupalayam Taluk, Tamilnadu. "Paper presented at IWMI - TATA water policy programme, 5th Annual Research meet", Institute of Rural Management, Anand, Gujarat.
18. Ratna Reddy, V. & Bhagirath Behera (2006) "Impact of Water Pollution on Rural Community—An Economic Analysis", *Ecology Economics,* Vol. 58, Issue 3, June.
19. Srinivas, Mudrakastra, Jain, Sheth and Srinath, J. (2006) "Unclogging the Khari River in Ahmedabad—Stake holders come together to halt pollution", *Economic and Political Weekly,* Vol. 41, No. 7, February-18, pp. 587.

20. Kunvar P. Singh, Dinesh Mohan, Sarita Sinha, Dalwani, (2004) "Impact assessment of treated/untreated waste water toxicants discharged by sewage treatment plants on health—Agricultural and environmental quality in the waste water disposal area", *Chemosphere,* Vol. 55, issue 2, April, pp. 227-255.

21. Phiri, O., Mumba, P., Moyo. Z., and Kaderva W. (2005) "Assessment of the Industrial effluents on water quality of receiving river in urban areas in Malawi", *Science and Technology,* Vol. 2, No. 33, pp. 237-244.

22. Chellappan Sekar (2003) "Effect of Industrial Effluents on ground water and the consequent impact on Economic, Social, Environmental, Governance", Paper presented at National Seminar on Water, Centre for Economic and Social Studies, Hyderabad, July 30,. 31.

23. Nagaraj, N., Chandrasekar, H., and Mangala, K.P. (2003) "Groundwater Pollution induced Externalities in Vrishabavathi River basin in Bangalore Hunter Lands : A Study in Peninsular India", Paper Presented at Information Support Sustainable Water Management: From Local to Global Level, Netherland 15-18 September.

24. Malik S. Negi (2003) "Impact of distillery effluent in water quality and phytoplankton in river saving", *Indian Journal of Environment and Science,* Vol. 7, No. 2. pp. 163-167.

25. Jothimani R. Baskaran B. (2002) "Effluents of dilution and dynamics of physical factor during factory Effluent irrigation". Environmental Monitoring, Vol.12, No.4, PP225.

26. Ahamed A. Alam (2003) "Physio-chemical and toxicological study of Industrial Effluent in and around Delhi", *Chemical Environment Research,* Vol. 12, No. 1 & 2, pp. 5-13.

27. Udaya Suriya and Prabhu P.C. (2003) "Impact of treated paper mill effluent irrigation on quality of soil and ground water", *Madras Agricultural Journal,* Vol. 90, No. &, pp. 540-541.

28. Patel, K.P. and Pandya, R.P. (2003) "Sustainability of Industrial Effluents for irrigation around Bharuch and Ankaleshwar Industrial Zone of Gujarat", *Pollution Research,* Vol. 22, No. 22, pp. 241-245.

29. Bhagirath Behara, Ratna Reddy V. (2002), "Environment and Accountability – Impact of Industrial Pollution on Rural

Communities", *Economic and Political Weekly*, Vol. 37, No. 3, January.

30. Fazeli, M.S., Khesravan, M., Hosini, S., Satyanarayar, P.N. Sathis, "Enrichments of Heavy metals in Paddy crops irrigated by paper mill effluent near Nanjangud Mysore Dt"., *Environmental Geology*, Vol. 34, No. 4, June 1998.
31. Rao, A.V., Jain, A.V., Gupta, I.C. (1993) "Impact of Textile Industrial effluents on Agricultural Land—A case study", *Indian Journal of Environmental Health*, vol. 30, No. 2, pp. 132-138.
32. Yadav, B.K. Lourdraj and Paul Sebastian (1998) "A Review on Industrial effluents use in agriculture—Problems and Prospects", *Crop Research*, Vol. 31 (2), pp. 183-191.
33. Aparna, S. & Anandavalli (1984) "A study on H_2O pollution by tanneries in Dindigul", *Indian Jounal of Pollution Control*, Vol., No. 2.

2

Water Pollution by River Noyyal

Water Pollution in Tamilnadu—A Glance

Tamilnadu geographic area consists of 17 river basins, a majority of which are water stressed. There are 61 major reservoirs about 40,000 tanks and about 3 million wells; that heavily utilise the available surface water (17.5Bcm) and ground water (15.3 Bcm).[1]

There are more than 3000 industrial units in Tamilnadu which have been classified under the highly Polluting (or) 'red' category. The total effluent generated is about 6 lakh litres per day, of which more than 5 lakh litre (85 percent) is generated by large industries. About 400 units discharge directly into rivers. Of particular concern are the nearly 1000 tanneries which are located in Vellore, Kancheepuram, Dindigul and Erode districts. The effluent has caused serious problems in number of districts. The Industrial pollution had led to rapid deterioration of the three main rivers of Kongumandalam (Erode, Coimbatore and part of Karur districts) with salinity touching upto 1700 mg per litre. These rivers namely Noyyal, Bhavani and Amaravathi once enriched the fertility of the land in their basins, but now polluting not only the drinking water source but lands and tanks as well[2].

The highly contaminated Bhavanai and Amaravathi rivers were polluting the waters of the Cauvery which irrigated about 64 percent of the agricultural land in the state. The three water starved districts were among the districts that recorded less rainfall compared to other districts. Similarly the effluent of Palar has caused serious environmental problems on the basin. A study carried out by the Loss of Ecology Authority, Govt. of India, revealed that the tannery industries have adversely affected 15,164 ha. of agricultural Land in Vellore district and 2,005 ha. in Dindigul district[3].

There are five main industrial complexes in Tamilnadu, Manali, Ranipet, Cuddalore, Mettur and Tuticorin which have chemical, petro-chemical and other industries. These complexes have environmental hotspots. There are Cement units, distilleries, Sugar, Sago, Paper, Dairying, Chemical and Fertilizers, Mining industries, mineral processing industries and a variety of other industries, which are water consuming and also generate large quantities of effluent.

Some of the industries have provided the treated effluent for irrigation with some degree of success. According to the state environmental rules, Industries cannot be set up within 1 km of a river (or) water body. However, the effluents often flow through open drains and reach the rivers along with sewage and sullage from municipalities and settlements has also increased tremendously due to piped water supply and in contaminating river, lake, tanks and groundwater.

All the industries discharging effluents are regulated by the Tamilnadu Pollution Control Board (TNPCB). They have met this effluents standards fixed by the board. Industries pay a cess based on their water consumption to the TNPCB. The state govt. has enacted acts for water pollution prevention and control. The systems rely almost exclusively on the downstream control by fixing effluent standard for the discharge of effluents in to water bodies. Regulations were framed to monitor and control the discharge of effluents from each industry and specification were laid down for the quality of effluents to be discharged in land (or) into water bodies after treatments. The government has taken several measures to prevent and control pollution of waterways. The government has passed orders banning the operation of highly polluting industries within 1km form the embankment of river, and reservoirs. The government had also passed orders imposing total ban of setting up of any of highly polluting new industries within 5km from the rivers of Cauvery and its tributaries, Periyar, Palar, Vaigai and Tamirabarani[4].

The River Noyyal

The Noyyal river originates from the Vellingiri hills in the western Ghats and flows through Coimbatore, Erode, Karur districts with its catchments in seven taluks, viz. Coimbatore, Tiruppur, Avinashi, Palladam, Kangeyam, Perundurai and Dharapuram and covers a total area of 0.35 million hectares with the distance of 180 kms. The boundary of this river is north Latitude of 10.54′ – 1.193′ and with eastern longitude of 76.39′-30 and 77.5′ and 25.[5] The river has the high terrain in the western periphery and flows on the plain and flows in the central and eastern parts. The river is also termed as Western Kailayam in Western Ghats and it ends in the river Cauvery at Kodumudi. Throughout its distance on both sides of the banks of the Noyyal River more than the 100 villages are situated, and it was the major source of irrigation, drinking water and other activities of the people living on both sides of the river and even for people beyond 3 km of river.

The river is flowing in a periodic manner and it receives copious water during North East monsoon from September to November and it remains dry state for six to eight months. The river being a seasonal one, fed only by the monsoon. The water-flow is moderate for a short period during the monsoon season. Occasionally flash floods occur after, in case of, heavy rains event. There are seven major tributaries, all originating from first (or) second order streams in the foothills of the Western Ghats. Floods are common during the rainy season due to steep slopes in the upper part of the catchments. Rainfall in the basin is highly variable due to the orographic effects of the Western Ghats. The mountain forms a rain shadow area over the plain which consequently has a dry climate. The western and upper reaches usually receive more than 3000mm annually during the south west monsoon where as the eastern part of the basin receives an annual rainfall of 600mm[6].

The Noyyal has a long illustrious history, as a river influenced by man, which is indicated by the many tanks and canals. Several civilizations have flourished around its banks throughout history. The present tank and canal system is very different from the one that existed before the start of urbanization. Today the basin appears to be fully exploited, having 23 ancient, 30 system tanks, 20 channels and two reservoirs, Orathupalayam and Athupalayam constructed for irrigation. It is also said that the water contains natural medicines and is therefore good for health.

Although Noyyal means "devoid of illness" but the river has become ill. It is today India's one of the most polluted river contaminating flora and fauna, ecology along the major stretches of the basin. The water originating from the hills is tasty and sweet but soon the quality changes, as much of the household wastes from the urban settlements are dumped into the river. Today parts of the Noyyal river basin are "Industrial Waste Lands" areas subjected to degradation as a result of large scale discharge of industrial effluents. The ground water close to and the water in the river, today unfit for drinking, sanitation or even irrigation in these areas.

Next to Chennai, Coimbatore is the most industrialized district in the state. The City is commercially vibrant, trade and commerce are the characteristics of the district. The textile industry, engineering industry, handlooms etc., have contributed in the small measure, towards stabling the industrial base of the district economy. The city Coimbatore, has been rightly called the Manchester of South India and could be compared with Bombay the Manchester of India. At present 256 medium and Large Scale Industries and more than 50,000 Small Scale Ir.dustries concentrating, textile, coir, chemical, food processing, etc[7].

Rapid Industrialization of Coimbatore district has resulted in the discharge of effluents/waste water into the rivers flows in the district viz. Noyyal and Bhavani[8].

A major part of the district falling around downstream of Noyyal river where the units of textile industry, which comprises bleaching, dyeing, printing and sizing emanate most of the contamination and pollution hazardous. Although the city of Coimabatore, a major town located the river Noyyal upstream of Tiruppur, has a very good sewage network according to the drainage pattern of the river but no waste water or effluent is released from the city into the river unless the taluk located within the limit of Coimbatore, overflows. Thus the entire stretch of the river Noyyal remains dry and clean until it receives the effluent at the down stream of Tiruppur.

Tiruppur, a major hosiery centre in south India has become a serious issue because it has had severe impact of water pollution along with the down stream area by contaminating the river Noyyal.

Tiruppur a small town, about 50km from Coimbatore, is known as one of the well known place in hosiery and knitwear manufacture in international market. Since 19th centuriy Tiruppur has its association with cotton trade. The first textile manufacturing units was established in the town in 1935[9]. It has entered into the international trade in 1980's with the manufacture of cardigans, jersy, pullovers, ladies blouses, skirts, nightwear, sportswear and industrial wear. Small scale garments manufacturing export had developed in Tiruppur rapidly in the 1980s and 1990s, using locally grown cotton in a small scale textile centre mostly from the informal sector. This growth was fuelled by a parallel decline in agriculture, whereby most of the farmer turned out for their opportunities which led them to invest in textile processing units.

Thus most of the firms in Tiruppur sub-cluster are typically small in size, and more than 90 percent are considered as small scale industries.

The main activities in industrial sector here are spinning, weaving, knitting, bleaching, dyeing, printing and allied works. The majority of bleachers and dyers around 78 percent are linked to the overall knitwear cluster through the job-work system, while 22 percent are independent producers. There are essentially three types of firms in the industry viz. direct exporters, indirect exporters and job-workers[10].

Table 2.1

Export Performance of Tiruppur Textile Industry

Year	*Qty. (Million Pcs)*	*Qty. in % of India's Knit wear Exports*	*Value (Million US$)*	*Value in % share in India's Knitwear exports*
1995-96	257.4	47.87	544.3	38.29
1996-97	294.3	46.54	604.2	37.77
1997-98	338.5	49.63	616.8	37.91
1998-99	368	48.52	689.7	36.58
1999-00	410.4	49.58	761.6	37.28
2000-01	372.4	51.87	718.6	40.30
2000-02	344.8	52.83	666.7	41.78
2002-03	370.4	54.57	793.5	43.62
2003-04	385.8	65.26	832.7	43.05
2004-05	396.4	75.2	848.6	47.20

Source: Apparel Export Council, Tiruppur.

Tiruppur has become an important textile cluster in India both for overseas market and the domestic market. It contributes around 70 percent of export knitted garment production in our country. In textile sector, India has the largest potential and Tiruppur has been identified as one of the small scale industrial cluster among the 358 clusters by UNIDO[11]. It has 2500 knitting and stitching units, around 800 dyeing and bleaching units, 300 printing units, 100 embroidery units and other 200 units catering to compacting and raisins. Buyers from around 35 countries visit Tiruppur regularly. This small town annually contributes about Rs 3600 crores in foreign exchange earnings to our country besides an earning matching (or) surpassing the above figure to cater domestic market.

The dramatic improvement in Tiruppur export market can be attributed to the disbanding the garment manufacturing in western countries due to environmental pollution and high cost of labour. The export of knitted garments from Tiruppur started to grow very rapidly around 1985 and in the 1990's the annual growth rate was 50 percent[12]. In hosiery exports, dyeing industries play critical role. It promotes primary and secondary employment opportunities, generate income to many people and also serves to the needs of the export earning of our government. It also provides breeding ground for allied industrial development, and increased competitiveness through public-private partnership.

During the last two decades particularly after trade liberalization global shifts have occurred in textile production and trade form the traditional production centers in the developed countries for the newly emerged centres in developing nations. Besides textiles, some of the other polluting industries have also migrated from developed countries to developing countries. One of the reasons attributed to this shift is the tight environment regulations and enforcements in developed nations when compared to the developing countries. Since the citizens of poorer countries demand only low level of environmental quality there is a considerable cost advantage in manufacturing the products through low pollution abatement cost. It is understood that the inefficient pollution management practices in Tiruppur as well as the lenient approach of the Pollution Control Board (PCB) has reduced considerably and provided a cost advantage to the exporters. In addition to this during the 1990's some western importers extended into agreements with the key garment manufacturers in Tiruppur to manufacture products in different brand names. At present the importers and consumers of Tiruppur garments are more concerned about the product price and quality and not the environmental issues associated with the manufacturing. Hence there is no environmental pressure on the exporters of Tiruppur from the international market.

Further the flux of outsiders from other parts of India to Tiruppur in pursuit of their business interests, and their ability to command the market with low levels of capital stock and capital intensity of production, have also significantly contributed to competition within the local community. The ability of outsiders to impart potential competition with the local community with a minimal capacity of production (measured by the ratio of the amount of capital that the company owns to its production) is attributed to the availability of job-workers and use of indirect exporters for decentralisation in the production process. This finally leads to large variations in the capital intensity of production in the major populations of direct exporters. This competition within the local production units has created pressure on the need for upgradation and modernisation to meet the needs of the local as well as international markets[13].

Irrespective of pressures on local and international markets, Tiruppur is poised for a spectacular growth, owing to the availability of raw material at lower prices, cheap labour, adoption of modern production technologies, and above all, the hard work of every one involved in the industry. Despite swifts in economic frontiers associated with fluctuations in the process of raw materials, it is interesting to note that the industrial growth in Tiruppur remains enabled. The success of Tiruppur lies in the efficient distributed network of small units, which shares a large export order into smaller tasks like a biological organism, and the whole town works like a decentralized factory for the global economy.

Environmental Problems in Coimbatore Regions

Textile production particularly, dyeing and bleaching are water intensive and could generate large quantities of effluents. Tiruppur is in a dry water scarce region, and the rapid expansion of the textile industry has taken place in an unplanned manner with no associated development of supporting infrastructure (or) institutional capacity. As a result, the growth has led to the depletion of ground water reserves and a serious detoriation in environmental quality of both surface and ground water. Typical water consumption in Tiruppur is around 200 to 400 litres/kg of finished product compared with the international norm of 120 to 150 litres per day[14].

It was understood from the available studies these units extract considerable quantity of ground water from the peripheral areas and discharge the effluent without adequate treatment. In the earlier years, industries met these water requirements by pumping water from the Noyyal River of from open wells. But the detoriation of surface and ground water quality in Tiruppur has compelled the industrialists to transport water from the peripheral areas to the city through tanks. A study by Balaji[15] (2008) suggested that, annually the textile industries in Tiruppur alone utilize around 28.8 billion litres of ground water. This is around 0.1 percent of present total water potential, 0.1 percent of water demand of Tamilnadu State, and 0.005 percent of world wide withdrawal of ground water.

Heavy utilisation of ground water has severely depleted the ground water sources in Tiruppur. Thus currently Tiruppur is faced with severe water scarcity. The water needs of the textile wet processing industries are met by the supply of water through lorries, taking up water from open and bore-wells in the surrounding Taluks viz. Avinashi, Palladam, Annur and Kangeyam. Roughly between 2000-3000 Lorries with a capacity to transport 10000 to 12000 litres/trip are plying around 7 to 10 trips daily to supply clean water for the textile wet processing.

At present 866 dyeing units engaged in dyeing and bleaching activities[16]. Most of these dyeing units are small industries and most of them undertake their order on job-work basis. These job-orders are mostly from direct exports, which opt to share the workload in order to meet the deadline, rather than facing cancellation of the export orders in the event of labour unrest and unfavorable weather conditions.

Out of the total water requirement, around 80 percent is daily transported from nearly villages and supplied by Lorries to the textile wet processing units in Tiruppur. The fresh water supplied by the Lorries is mixed up with the locally available ground water for dyeing process in order to save water costs. As the ground water has higher (TDS) Total Dissolved Solids (around 2500-3000 mg/L) use of this water in textile dyeing process has further enhanced the TDS value of the discharged effluents and it becomes difficult to treat the effluents to achieve desired levels of TDS for discharging as per the norms and standards of Pollution Control Board. According to the Tamilnadu Pollution Control Board, an estimated 83 million litres of effluent water is discharged from the units.

Most of the bleaching and dyeing units in Tiruppur are located in clusters along with the banks of Noyyal River with which they discharge the effluents. The river is natural drainage courses that can only carry waste in monsoon period. During the rest of the years, they only carry industrial effluents that stagnates the river bed and percolate into the ground water. As a result, the ground water quality around the cluster of bleaching and dyeing units is polluted to such a level that it is unfit for domestic, industrial and agricultural activities. The bleaching and dyeing units are mainly responsible for polluting the environment of Tiruppur and the down stream areas. Clean water is the critical requirements of these dyeing units, as the quality of the water largely affects the quality of dyeing and bleaching. While these industries use clean water for processing the chemicals used by these industries viz., wetting agents, soda ash, caustic soda, sodium hypochlorite, bleaching powder, acids, dye stuffs and soap oil make the water very harmful to the environment.

A total of eight Common Effluent Treatment Plants (CETP) are in operation at Tiruppur town with design capacities ranging between 1.6 to 10 billion litres which generally receive the waste water from these units. At present a total of 281 units are discharging into the CETPs having a total design of 42.05 million litres, whereas the rest having their Individual Effluent Treatment Plant (IETP). Almost all the CETPs are operating at their design capacities. The transport of water to the CETP s through the pipelines and no tankers are allowed to carry the waste water to CETPs. The CETPs have been designed on physio-chemical treatment process. As such the treatment systems have not been designed to contain total dissolved solids from the waste water. On the contrary due to the mixing of lime, ferrous sulphates, coagulant acids the TDS levels increase in the effluents leaving the CETPs.

Among the total effluents generated by these industries, the CETPs able to treat only 53 percent of the effluents, while the rest are discharged into the river without any treatments. Though the CETPs are successful in removing colour of the TDS solids of the discharge to the permissible limit of 2100 mg/l for the reason is that good dyeing requires high concentration of salt. Thus discharge of high TDS effluent into the river also damaged its purity. More over Biological Oxygen Demand (BOD) in the raw effluent cannot be treated to the desired BOD concentration for river disposal. Thus CETPs are providing only partial treatment and requires additional units for treatment of TDS and BOD.

Thus it clearly reveals that the economic base of the knitwear industry in Tiruppur region along with the magnitude of environmental damage and the inefficiency in existing pollution management. In order to solve this problem, it was assumed that installation of effluent treatment plants, but this is less possible as these plants are not designed to remove TDS. However pollution is not the only concern of our industrial sector but the industries too. Since the water resources are extremely scarce in Tiruppur region, extraction and transporting a large quantity of fresh water from distance and then the discharging the entire waste water into the river is not a sustainable management practice.

There are number of Science studies were carried out to analyses the pollutants of textile units in Tiruppur and their impact on ecology and flora and fauna. The ground water studies indicated that drinking, sanitation and even irrigation is critical in Tiruppur and the downstream areas. Few of them are herewith presented.

A study by Nelliyat[17] (2005) showed that open wells and bore wells in and around Tiruppur exhibit high level of TDS (ranging 3000mg/1 litre 11,000 mg/ltr) and chloride (ranging 200mg/1 ltr 5000mg/1 ltr) due to industrial pollution and these values are much higher than the background level for this region.. During 1980 to 2000, the cumulated pollution load discharged by the Tiruppur units comes to 2.35 million tonnes of TDS, 1.31 million tonnes of chloride, 0.13 million tonnes of sulphate, 6.098 million tonnes of Total suspended solids, 0.09 million tonnes of Chemical Oxygen Demand, 0.03 million tonnes of Biological Oxygen Demand and 0.001 million tonnes of oil and grease. The total damage cost is the drinking water sector is about US $23.8 million.

A comprehensive study was made by the Soil Survey and Land Use Organisation as part of the project study made by the Madras School of Economics during 2001 to 2002 to assess the environmental impact of industrial effluents in the Noyyal river basis. Both water and soil were surveyed to investigate quality of water and soil for agricultural use. The study covered a stretch of 5 to 6.5km on either side of the river from Vellingiri to confluence point and was divided into four zones. The study indicates nearly 33 percent (82 samples) indicated that it was not suitable to grow crops and 42 percent (106 samples) were in the critical category. Around 23 percent of the area (65 soil samples) was not used for cultivation.

Senthilnathan[18](1998) took surface water samples at five locations in Noyyal river, the first one was upstream of Tiruppur and the last one is the Orathupalayam dam. The water sample, except for the first and last sample was dark blue to black and had a bungent smell. It was mainly by the unspent dyes that cause the color. The color impedes light penetration and these effects the biological activity. Colour removal from effluents is a major environmental problem because the chemical treatment results in huge quantities of sludge.

Azeez[19] (2001) summarises his study and showed that in many places of the downstream of Tiruppur, the water samples was about 800mg/l and total hardness about 400mg/l. The water was very turbid because of the contaminated materials generated in the industrial units. The total solid was above 10000 mg/l and TDS has reached almost the same level. The sodium concentration was very high and concentration of chloride reached upto 4360 mg/l in some surface water samples. The Dye materials settled from the effluents has changed the sediment colour since components in the sediment system absorb the dyes and chemicals. This is also the case with terrestrial soil flooded by polluted river water or irrigated with this water.

A study carried out by Public Works Department (PWD) on "Environmental status of the Noyyal river basin" on 2001. The surface water samples were collected at 28 locations. The study shows that EC and TDS values were relatively low until the river reached Tiruppur. The TDS values were highest in the Tiruppur area and remained high up to Orathupalayam Dam. From this dam, TDS showed decrease, since the water stagnated in the dam and was not often released, and probably because the Lower Bhavani Project (LBP) canal crosses the Noyyal river downstream of orathupalayam dam.

Sankarraj[20] *et al,* (2002) studied on soil quality in the area around the Orathupalayam dam showed that most of the soils are naturally non-alkaline. However as a consequence of the pollution, the soils have started to become increasingly alkaline. There has been an increase in both soil pH and EC. Other effects that have been noticed and described are for example soil colour, structural changes and decreased crop yields.

The research department of Zoology of Erode Arts College conducted a biological study in 1998 on the impact of pollution on aquatic life. The study had found that the plankton spices diversity was comparatively low from Sulur to Orathupalayam. There was a high number of plankton and fish diversity downstream of the dam because of the diluting effect of seepage water from the LBP canal. Indicators such as liver glycogen and blood glucose level in fish confirmed that the water in the dam is more toxic to fish than water in further downstream. In 1997 a mass fish mortality occurred in the reservoir. Floating dead fish could be seen near the dam site and posed a serious threat to human health.

Govindarajulu[21] (2003) analysed the health status of the people residing on the Noyyal river basins to identify the impact of the water on health status, 3 health camps were conducted at 3 centers on different dates, 20 villages of which 1120 villagers attended the camps. It was identified that 31.65 percent of the patients were suffering from skin allergy problems, 22.56 percent of the patient held respiratory problems, 13.13 percent of general allergy, 10.44 percent had joint pain, and 7.07 percent ulcer problems were scanning, among the villagers who attended the medical campus. It was medically accepted that the polluted water had significant influence on these diseases. It was assessed that one-fourth of the villagers has any one of the listed of these diseases. Most of the identified patients accepted that they were using either the river water (or) well water or both.

Palanivel and Rajagopalan[22] (1999) study on the present status of the Noyyal River provides a detailed physio-chemical analysis of surface water quality in Noyyal River at different zones and compared the parameters in summer and winter. The water was collected from 28 spots extending over 101 km of the Noyyal River, of them zone-1 was upstream area of Tiruppur and zone II was downstream area of Tiruppur. The result of the study showed that *a)* compared to zone III and II, zone I is less polluted area, *b)* self-purification capacity of the river is a reason for the marginal reduction in pollution downstream, *c)* since the concentration of TSS and TDS is very high in Tiruppur and downstream, the river water is not suitable for any purpose including irrigation, and *d)* the existing moderate flow, during the monsoon is not sufficient to wash away the pollution.

An environmental damage cost assessment study by Appasamy[23] showed that the estimated overall annual damage cost in the agriculture sector was Rs. 35.29 crore and the capitalized value at 12 percent is estimated to be Rs. 235.54 crores. The annual loss of value (productivity value) is the fishery sector was Rs.0.15 crore and the capitalized value at 12 percent is approximately Rs. 1.25 crores. In the areas of drinking water sector was estimated to be about Rs.15.05 crore per annum and capitalized value at 12 percent is Rs. 109.03 crores. In summary the total social cost of pollution in agriculture, fishery, drinking were estimated to be Rs. 48.41 crores per annum. Similar study was conducted by Madras school of Economics reported that the annual fish catch in the Noyyal river declined 2174 kg (94.95) to 540 g in (1997-1998) and the corresponding value reduction was Rs. 27175 to Rs. 6210 in current prices. The fisheries activities in eight system tanks of the river (between Tiruppur and Orathupalayam) were highly affected. The panchayat used to auction the fish catch in five of the tanks, which had high values. The auction rate is reduced in all Panchayat and the total annual reduction between the year 1999-2000 was Rs. 2,57,000 in current prices.

In 2004 a team of students from the Jaivabai Municipal Girls Higher Secondary School in Tiruppur had conducted a study on the area around the Orathupalayam dam. The team covered eight villages on a 100km stretch along the Noyyal, collected water samples from 10 locations near the dam, tested them and got them counter-tested. It was found that the water was unfit for cultivation. Infact it was found unfit for use even in construction. The water was also found to have higher than permissible level of lead and cadmium.

Legal Proceedings of River Noyyal

The environmental pollution of the Noyyal River has evoked legal proceedings against the pollution. They are discussed below.

The Loss of Ecology Authority (LEA) appointed by the Madras High Court green bench to assess the damage caused by ground water pollution by the Tiruppur dyeing and bleaching industries, has asked to pay the pollution Rs. 104 crore as the compensation of the farmers of the Noyyal river basin at the adjoining Erode district for agricultural crop loss suffered by them in the vicinity of the Orathupalayam check dams. The cash compensation to the Orathupalayam farmers has been made out on the basis of damage assessment study done by a team from the Anna University's Department of Environmental Science engaged by LEA. The compensation has been made for the loss incurred by the farmers under the Orathupalayam dam ayacut for a four year period between the year 1999-2002[24].

In December 2004 the LEA ordered that the dyeing and bleaching units of Tiruppur should pay Rs.24 crores as compensation to farmers for polluting the ground water by discharging effluents.

In August 2006 the Madras High Court appointed the committee in case filed by the Noyyal River Aya cuttars (Beneficiaries of Noyyal river), who alleged that effluents from dyeing units in Tiruppur had polluted the riverbeds and then their agricultural activities. The committee comprising the former Engineer in chief of Public works Departments and the collectors of Erode and Coimbatore. The court has instructed the two collectors to provide assistance in the cleaning operations of Orathupalayam dam. The committee also deliberated on the fine paid by the polluting dyers in Tamil Nadu[25].

In the second week of July 2005 the High Court of Chennai ordered that all dyeing and bleaching establishments who have not compiled with the norms of TNPCB and have not achieved zero discharge system so far should be closed with effect from 19th July 2005. In order to show the bonafide of the industrialists towards establishment of the zero discharge system, the court ordered that 25 percent of the cost of the Reverse Osmosis unit should be to the supplier, and then the unit can seek the court's direction for revoking the closer order. The court insisted on seeing tangible proof of the owners sincerely of intent in the shape of demand drafts for the money paid as advance within few days. Most of the dyeing units had managed to present the proof, according to the sources. 493 dyeing and bleaching units had managed to pay Rs. 40-50 crores, while more than 100 units has escaped from court's order as they had already installed Reverse Osmosis plants[26].

In 1989, The Tamilnadu Government (Order No. 213) issued an order that banned the location of any industrial unit that discharges untreated effluent one km of river. Although many units were established in Tiruppur before this, and many of them had extended textile processing capacity significantly. Since then the government has not re-examined the issue even though a subsequent order extended this ban to a distance of 5km, dyeing units come under red category, meaning that they produce hazardous waste and therefore cannot exist within 5km radius. However all this presently existing dyeing and bleaching units within a radius of 5km were established before enacting this rule. As this rule was applicable only to the newly emerging units, the older units were allowed to operate within 5km radius of river Noyyal.

The recent order by the Chennai High Court that the industrial units in Tiruppur will have attain "zero discharge" by the deadline of July 2007 or shut down other operations. The High court order judgments requires all the dyeing and bleaching units in Tiruppur to setup individual effluent treatment plant (or) utilise the service of the common effluent treatment plan to achieve this. The order therefore requires the units to recycle their waste water. For recycling the units will have to put up three stage Reverse Osmosis Processing plants in place, which cost anywhere to the tune of Rs.2.25 crore for medium size units handling around 1,00,000 litres of waste water per day. The recycled waste water that the units get can be in the range of 50 to 90 percent of the waste water[27].

The Public Accounts Committee (PAC) of the Tamilnadu Legislative Assembly proposes to recommend a Rs. 800 crore of scheme to dispose treated effluents from Tiruppur into sea. The committee has visited various places in the district and admitted that the water from the Orathupalayam reservoir had turned out to be as storage for effluents from Tiruppur. The entire agricultural land around the reservoir has been affected and the water from the surrounding well is not suitable for drinking. They reported that number of patients in Kanchipuram Cancer Institute were from this region[28].

REFERENCES

1. Government of Tamilnadu (2006) "Environmental Concern of Tamilnadu/Agriculture" State Environment Report. Department of Environment, Chennai.
2. Government of India (2005) Tamilnadu Development Report. Planning Commission of India, New Delhi.
3. The Editor (2001) "Industries polluting various Rivers in Tamilnadu", *The Times of India* (News Daily) November, New Delhi.
4. *Op. cit.* Govt. of Tamilnadu.(2006).
5. Centre for Science and Environment, New Delhi, website available at www.rainwaterharvesting.org
6. Krishna Furn (2004) Effects of dyeing and bleaching industries on the area around the Orathupalayam Dam, Southern India, http://www.w-program.nu/.
7. Profile of Coimbatore District (2005) District Collectorate Office, Coimbatore.
8. *Ibid.*
9. Banarjie and Munshi, K. (2003) "How efficiently is capital allocated? Evidence from the knitted garment Industry in Tiruppur", Bureau for Research in Economics Analysis and Development, working paper No. 004 Massachusetts Institute of Technology.
10. Thomas Jacob, Jayabal, Azariah (1997) "Environmental perception of Textile Industrial Pollution in Tiruppur, India" *Eubios Journal of Asian and International Bio-ethics*, vol. 7, pp. 162-165.
11. Eswaramoorthi & Dhanabal (2005) "Textile effluent and waste water management in Tiruppur, India—Need for Government Aid" Report prepared by Environment with People's

Involvemenι & Co-ordination in India, (NGO), August.

12. Crow, M.T. (1999) "Successfully Adjusting the Environmental Regulation; The small firm cluster of Tiruppur; India", Master in City planning, Massachusetts Institute in Technology, pp118.

13. Sakthivel A (2004) "Tiruppur Knitwear Export Cluster—India ITC Executives Forum on "National Export Strategies—competitiveness through public private partnership success and lessons learned", Montrex, Switzerland, September 26-29.

14. *Op.cit.*, Dhanabal.

15. Balaji, S. (2003) *State Environment Report of Tamilnadu*, Chennai.

16. Government of Tamilnadu (2007) "Unpublished record of Tamilnadu Pollution Control Board, Tiruppur Branch.

17. Prakash, Nelliyat (2005) Industrial growth and water pollution in Noyyal River Basin, Working Paper Series, Madras School of Economics, Chennai.

18. Senthil Nathan, S. (1998) "Environmental impacts of dyeing and bleaching and dyeing industrial effluents on abiotic and biotic components on river Noyyal", Paper presented at the Seminar on Impact of Industrial effluents, Sewage and agricultural practice on water quality, by Public Works Department, Coimbatore, August.

19. Azez, P.A. (2001) "Environmental implication of untreated effluent from bleaching and Dyeing, (ed). Senthilnathan, Eco friendly technology for waste minimization in Textile Industry, Centre for Environment Education, Newdelhi.

20. Sankar raj, L. Subramanyam T.P. Siddhamalai A & Faroque Ahamed .N. (2002) "Quality of Soil and Water for agriculture in Noyyal River basin. Tamilnadu" Soil Survey and Land use organization, Department of Agriculture, Coimbatore.

21. *Op.cit.*, Govindarajulu.

22. Palanivel & Rajagopalan (1999) "The present status of the Noyyal river" Proceedings of the workshop on Environmental Status of Rivers in Tamil Nadu", Bharathiar University, Coimbatore.

23. Pual P. Appasamy, Prakash Nelliat (2007) "Compensating the loss of Ecosystem services due to Pollution in Noyyal

River Basin, Tamilnadu, Working paper series 14/2006, Madras School of Economics, Chennai.

24. The Editor (2004) "Tiruppur dye sector in pollution—told to Folk out Rs. 104, on compensation", *The Hindu* (News Daily) March 17, Chennai.
25. Satyamoorthy (2006) "Tiruppur effluent discharge—PAC marks Rs 800 on scheme", *The Hindu* (Daily News) November 2, Chennai
26. Sridhar (2005) "The farmers plight", *Front Line* (News Magazine) vol. 22 No. 17, August.
27. Gaurav Dwivedi "Tiruppur water supply and sewage project". *News for Sustainable Development*, News, No. 17, Bangalore.
28. The Editor (2006) "High Court Monitory Committee visits Orathupalayam Dam", *The Hindu* (News Daily), August 18, Erode.

3

Major Polluting Rivers in India—A Glance

The present section reviews the major polluting rivers in India. Meanwhile explaining of all the polluting rivers in India is formidable. An attempt is made to review few of the major polluting rivers and the causes of pollution.

The Ganges River

Winding 1,560 miles across northern India, from the Himalaya Mountains to the Indian Ocean, the River Ganges is not a sacred place: it is a sacred entity. Known as Ganga Ma—Mother Ganges—the river is revered as a goddess whose purity cleanses the sins of the faithful and aids the dead on their path toward heaven. But while her spiritual purity has remained unchallenged for millennia, her physical purity has deteriorated as India's booming population imposes an ever-growing burden upon her. The river is now sick with the pollution of human and industrial waste, and water-borne illness is a terrible factor of Indian life. But the threat posed by this pollution isn't just a matter of health—it's a matter of faith. Veer Bhadra Mishra, a Hindu priest and civil engineer who has worked for decades to combat pollution in the Ganges, describes the importance of protecting this sacred river:

The tremendous life that the Ganges supports is also the source of its greatest threat is pollution. The majority of the Ganges' pollution is organic waste—sewage, trash, food, and human and animal remains. Over the past century, city populations along the Ganges have grown at a tremendous rate, while waste-control infrastructure has remained relatively unchanged. Sewage systems designed near the turn of the 20th century today do little more than channel waste into the river. Around 300 million gallons of waste go into the Ganges each day, and the effects are stunning: recent water samples collected in Varanasi revealed fecal-coliform counts of about 50,000 bacteria per 100 millilitres of water, 10,000% higher than the government standard for safe river bathing. The result of this pollution is an array of water-borne diseases including cholera, hepatitis, typhoid and amoebic dysentery. An estimated 80% of all health problems and one-third of deaths in India are attributable to water-borne diseases.

The sacred practice of depositing human remains in the Ganges also poses health threats because of the unsustainable rate at which partially cremated cadavers are dumped. In Varanasi, some 40,000 cremations are performed each year, most on wood pyres that do not completely consume the body. Along with the remains of these traditional funerals, there are thousands more who cannot afford cremation and whose bodies are simply thrown into the Ganges. In addition, the carcasses of thousands of dead cattle, which are sacred to Hindus, go into the river each year.

While industrial pollutants account for a smaller proportion of contamination in the Ganges, the health and environmental impacts of toxic chemical waste can be far greater. From the plains to the sea, pharmaceutical companies, electronics plants, textile and paper industries, tanneries, fertilizer manufacturers and oil refineries discharge effluent into the river. This hazardous waste includes hydrochloric acid, mercury and other heavy metals, bleaches and dyes, pesticides, and polychlorinated biphenyls—highly toxic compounds that accumulate in animal and human tissue. Runoff from farms in the Ganges basin adds chemical fertilizers and pesticides such as DDT, which is banned in the United States because of its toxic and carcinogenic effects on humans and wildlife. Damming the river or diverting its water, mainly for irrigation purposes, also adds to the pollution crisis.

River Yamuna

About 57 million people depend on Yamuna water, with a annual flow of about, 10,000 cubic meters (cum) and usage of 4,400 Cum (of water irrigation constitutes 96%) the river accounts for more than 70 percent of Delhi's water supplies.

Yamuna's pollution starts from Tejewala in the upper segment. Here two canals, the Western Yamuna Canal (WYC) and the Eastern Yamuna Canal (EYC) divert river waters. The WYC crosses Yamuna nagar, Karnul and Panipat before reaching the Haidarpur treatment plant (which supplies part of Delhi's water) receiving waste water from Yamuna Nagar and Panipat.

At Panipat, discharges from the Panipat Sugar mill and distillery are let out into a disused canal. Some times, when the effluents cross the dam, it results in a major increase in Biological Oxygen Demand (BOD) loads in the WYC. Central Pollution Control Board (CPCB) inspection report estimated that there were 1,00,000 cum of effluents in the disused canal, having a BOD level of 1380 mg/l.

Yamuna enters Delhi at Palla village 15 km upstream of Wazirabad barrage, which acts as a reservoir for Delhi. Delhi generates 1900 million litre per day (mld) of sewage against an installed waste water treatment capacity of 1270 mld, Thus 630 mld treated sewage enter the river everyday. From the Okla barrage, this is the exit point for the river in Delhi, the Agra Canal branches out from Yamuna. During the dry months, almost no water is released from the barrage to downstream Yamuna. Instead, discharges form the Shahadara drain join the river downstream of the barrage, bringing effluents from east Delhi and Noida into the river. The main problem lies in undetected and untreated pesticide residues.

The River Bhavani

Bhavani, the second largest river in Tamilnadu begins from Kerala's silent valley and flows into western Tamilnadu, covering a distance of 217 km before merging with the Cauvery. The basin drains an area of 0.62 million hectare, spread over Kerala (9%) Karnataka (4%) and Tamilnadu (87%). The main river courses through Coimbatore and Erode districts of Tamilnadu, before reaching the Cauvery at Bhavani town. About 90 percent of the river's water is used for agriculture, even as industries dot the sub basin at every point.

Pollution Problem

Industrial pollution together with domestic pollution has proved to be the base of the Bhavani threatening the survival of 6 million people on the bank. Although numerous industries are located along the path of the river, it is chiefly SIV industries Sirumugai, a multi project private sector undertaking, has completely damaged the lower, basin, while textile, tanneries and dyeing industries where have further downstream of SIV. About 38,000 cubic meters of effluents and waste water are dumped into the river. The problem manifests itself when pollutants collect downstream at the Bhavanisagar dam.

In addition to the above industrial waste and municipalities also dump sewage in to the river. Untreated sewage from Ooty, Coonur, Mettupalayam, Satyamangalam, Gobi, and Bhavani town are dumped into the river. At Mettupalayam, numerous slaughterhouses add their waste to the domestic sewage of the town. The Nilgiris mountain range is also an important contra for pesticide-dependent tea plantations. Pesticides from there tea estates find their way to Bhavani. Tea estates and coffee pulp houses add or estimated 1.5 mld of effluents to the river every day.

River Krishna

The river originating form the western flanks of the Western Ghats near Mahabaleswar. Krishna site fourth largest river basin in the country. It flows 1440 kms through the states of Maharashtra, Karnataka and Andrapradesh before confluence into the Bay of Bengal. The rivers Koyna, Varna, Panchganga, Dudhganga, Ghatapraba, Malapraba and Tungabadra join Krishna from the right bank; while the Yarla, Musi, Maneru, and Bhima rivers join the Krishna from the left bank. The river receives effluents and waste water from a number of large cities, including Pune, Satara, Kolhapur, Hyderabad, Karnool and Vijayawada. More than 500 important industrial units operate from the Krishna basin, 200 of which are large scale units. Sewage from Pune is choking the City's river Mula-Mutha, Musi, Krishna's tributary on the west Hyderabad, Andhra Pradesh is in a bad shape with nearly 350 mld of polluted water and sewage originating from the twin cities of Hyderabad and Secunderabad flowing into it.

Pollution Problem: The tributary of Krishna, Tungabadra was also damaged by the industrial pollution. Two units viz., Gwalior Rayon Silk manufacturing company Ltd. are located in on its banks of Dharwad disrict of Karnataka where the downstream villagers of the factories felt the effects of pollution almost as soon as the industrial units was commissioned. The river water became dark brown in colour and has a pungent odour. The pollution of Tungabadra affected 1,00,000 people in the sub basin. Instances of stomach ailments and skin rashes were recorded by the local administration as most villages used the river for drinking water, bathing, irrigation, fishing and water for their livestock.

But the worst affected were the fishing community. Regular fish kills have exhausted Tungabhadra's fisheries. In February 1984, a massive fish kill was reported in the Tungabhadra, downstream of the factories. The state government has failed to act despite the river's grim pollution scenario. Despite protests from the villagers and farmers, all that the state government has formed several committees to study the problems with no result on the ground.

River Godavari

The second largest river in India Godavari is often referred to as 12 Vriddh (old) Ganga (or) the Dakshin (South) Ganga. Pollution in this peninsular river is fast reaching unsafe levels. The Godavari originates near Triambak in the Nasik District of Maharashtra, and flows through the states of Madhya Pradesh, Karnataka, Orissa is just 80 Kms away from the Arabian Sea. It journeys to 1,465 kms to fall in the Bay of Bengal. Some of its tributaries include, Indravathi, Manjira, Bindiusara and Sabari. Some important urban centres on its banks include Nasik, Aurangabad, Nagpur, Nizamabad, Raja Mundry and Balaghat.

Pollution Problem: Like most other rivers, domestic pollution is the biggest pollution of the river Godavari, accounting for 82 percent of total pollution, whereas industrial pollution accounts for about 18 percent. Over half of the river basin (18.6 million hectare) is categorized as cultivable land. Most of the river's water is drawn for irrigation purpose. The problem of pollution in the Godavari River evolves around the tiny, Nagavakku river bed which joins the Manjira, a tributary of Godavari. Highly productive agricultural lands surround the rivulet. More than 150 small and medium industries and several large industries near the twin cities of Secunderabad and Hyderabad, release their effluents into the Nakavaggu rivulet.

However most of the blame lies with the 72 industries in the Patanchera Industrial Area that have been dumping their effluents into the river. Industrial discharge from such industries has severely affected public health, surface and ground water and agriculture in 22 villages in this are. The river water is heavily used for agriculture as it is the only available water source. However the river's water has turned the fertile soil toxic with heavy metals. Before industrialization, the lands crop yield was 40 bags of Paddy per acre and now only 10 bags of paady per acre. Toxic metals in the soil have contaminated the crops, penetrated animal milk, and affected human health.

Regarding the governments action in 1993, the Aurangabad Bench of Bombay High Court ordered two paper mills in Paithen to stop discharging effluents with the river and then Beed Zilla punished to supply drinking water for the affected village. Moreover, after prolonged agitation by farmers and pressure from citizen bodies, the district administration got the court to serve notice to 22 to villages establish individual Effluent Treatment Plants. (ETP's)

River Subarnarekha

The rain-fed of Subarnarekha originates 15 km south of Ranchi on the Chottanagpur plateau draining the states of Jharkhand, Orissa and West Bengal before entering the Bay of Bengal. The total length of the river is 456 kms and its important tributaries include the Rani, Kanch, Karseri, Kharkai, Garra and Sankh rivers.

Pollution Problem: The only streaks visible in the river are those of domestic industrial radioactive pollution. Subarnarekha rich resource basin has spelt doom for the basin between Mayurbhanj and Singhbum districts on the right banks of the Subarnarekha are the country's richest copper deposits. The proliferation of unplanned and unregulated mining and mineral processing industries has led to a devastating environmental degradation of the region. During monsoon, this exposed earth flows into the river increasing suspended solid and heavy metal load in the water, silting the dams and reservoirs.

Subarnarekha also has to bear radioactive waste that enters the river through seepage form tailing ponds of the Uranium Corporate of India at Jadugoda. It has three productive Uranium mines, all within 5 km radius. Jadugoda, Batin and Narwapahar. The tailing (or) effluents comprising radioactive products, which are mixed into slurry and pumped into tailing ponds. These ponds each covering each 160 ha of land and about 30 meters deep are situated between adjoining villages. These radiations pose the greatest threat to human health, as they harm living cells, often leading to genetic mutation, cancer, and slow death.

Subarnarekha is the life line of tribal communities inhibiting the Chottanagpur belt. Between 5000-6000 families of local tribal, including the fishing community of Dharas, residing on the river banks from mango in Jamshedpur to Bharagora have been affected by the river's pollution.

Government action has been all but absent in the basin. Although the National River Conservation Plan (NRCP) intends to clean domestic waste generated from Jamshedpur, Ranchi and Ghatsila, they are supposed to be tackled by respective state pollution control boards. Almost industries and mines in this basin fall in Jarkhand, the ones by checking pollution falls squarely on the state pollution control board which has a particularly poor record in this regard.

River Damodar

The 563 km long Damodar originates near Chandwa village in the Chottanagpur hills in Bihar, Palamau

District, it flows through one of the richest mineral belt in the world before draining into the Hoogly, about 50km south of Calcutta. In the upper valley area, mining and mine-based industries are the dominant economic activity. With low agricultural productivity, combined with the heavy mining activity in this area had made the valley vulnerable to soil erosion. More than 50 major and medium industries and over 400 industrial units dot the valley. The states of Bihar and West Bengal depend almost entirely on this area for their power requirement.

The Damodar and its tributaries drain almost the entire coal mining area under the Central Coalfields Ltd. (CCL) the Bharat Coking Coal Ltd. (BCCL) and the Eastern Coalfields Ltd. (ECL) all the three subsidiaries are the public sector Coal India Ltd., the CIL is poised to produce 370 million tones of coal by the turn of the century. Thus about 60 percent of the coal extracted from the area comes from large, open cast mines. These mines are serious sources of land degradation mechanical extraction of coal leads to the mining of both coal and rock since the machines can't distinguish between the two. The rock mixed just adds to the volume of waste generated the total volume of overburden, which is about 200 million cubic meters, is likely to be 500 million cubic meters by the turn of the century.

The people who live in the vicinity of the Damodar are the worst affected. The river and its tributaries are the largest sources of drinking water for the huge population that lives in the valley. The water from the river that the people drank is unfit for human consumption, with oil levels 40-80 times higher than the maximum permissible value & 0.03 mg/l. The urban population of more than 3 million in Bihar and West Bengal is supplied by the Damodar water after treatment with Lime and Chlorine. The large rural population and urban dwellers outside the industrial townships are benefited of clean water.

River Khari

The river Khari originates from the hills near Nandol, 20 km east of Gandhinagar, bordering Ahmedabad district Gujarat. This first order river joins the Meshwo River a tributary of the Sabarni near Vartha on the border of Kheda district. Meshwo meets the Sabarmati at Kheda 50km down the route. The 80km Kharcut Canal that begins from Raipur village was constructed more than 100 years ago. Providing irrigation support to 10,200 ha in 110 villages spread over 80km in Daskroi taluk of Ahmedabad and Mahendavad taluk of Kheda district.

For more than two decades the people in this belt have been suffering from land and water pollution. Health related problems include skin disease stomach diseases.

Owing to the absence of adequate and appropriate drainage system, the industrial estates of Naroda, Odhar, Vatvaa and Narol, let out sewage into the Khaarient Canal. Khari River and this canal have now become a bone of contention between the industries and the farming community in positional those who live on the banks. Ground water contamination resulting from the illegal release of untreated effluents in to the river is the serious problem. The people of the river bed are forced to drink this polluted water in the absence of an alternative source. Many families walk long distances to fetch comparatively better quality water from farm tube wells.

Haora River

Haora River is the lifeline of Agartala city because it fulfills the major demand of drinking water as well as water for other purposes of the entire population of Agartala city. It also fulfills the total demand of the families who reside near the banks of the river from Champaknagar to Bangladesh border area. Haora is one of the ten major rivers of Tripura. The river originating from the eastern side of the Baramura range flows westerly through the alluvial plains and passes by the southern embankment of the capital city of Agartala before finally flowing down into Bangladesh. The flow length of the river is 53 kms in Indian territory and basin area is 570 sq.kms. The annual flow of this river is 36,032 m^3. Agartala Municipal Council and Public Health Engineering Department uses the water of this river daily for community use. The intake wells are situated near Jogendranagar and Gandhighat.

The water of Haora River is the main source of drinking water for the people of Agartala and its surrounding areas from Champaknagar to Bangladesh border. Previously the river water was pollution free due to less anthropogenic interferences. Presently, the number of inhabitants on the banks of the river is increasing very fast. As the inhabitants of the river bank belong to the section Below the Poverty Line, they have to make kaccha latrines on the bank of the river and their excreta flow directly into the river. Due to this, the river water quality has deteriorated to an alarming condition.

A study by Tripura State Pollution Control Board showed that the water quality of the Haora River has confirmed that the pollution load of up-stream of Haora River is less as compared to down stream. The pollution load especially of Biological Oxygen Demand (BOD) and Total Coliform increases and Dissolved Oxygen (DO) decreases towards down stream. From the study it is revealed that untreated sewage, human excreta (approx.6654 Kg/day) are directly discharged into the Haora river everyday from 1095 numbers of kaccha latrines of 1145 households on the river banks. It is also seen that Bacterial contamination including pathogens and solid wastes from different sources like cremation ground etc. are discharged directly into the river in the Urban areas.

River Palar

The river Palar is one of the worst polluted rivers in Tamilnadu. Palar in Tamil means river of milk, as it was giving sustenance to people of these parts of Tamilnadu in those days. Now it has become 'Pazhar' i.e. 'river of waste' major part of the river has become cess-pool of effluents and sewage drain for hundred of tanneries.

North Arcot Ambedkar District in western part of Tamilnadu has carved out a niche in the map of India, not only for its leather exports, but also for the high rates of water pollution. Nearly 574 tanneries discharge their effluent in to the river Palar. The river Palar passing through the district had become a river with polluted under ground water and in the Vaniyambadi-Ambur belt on the western end, and Walajapet and Ranipet belt in the eastern side. A survey conducted by the Tamilnadu agricultural university of the pollution caused by tannery effluents in the districts showed that out of one lakh hectare in the 1000sq.km of tannery belt from Vaniyambadi to walajapet, more than 35,000 hectares of valuable agricultural land have become either partially or totally unfit for cultivation. The area paddy cultivation has been drastically reduced by 40 percent in the tannery belt.

The Tamilnadu Pollution control Board has succeeded to some extent in forcing affluent tanners to put up individual Effluent Treatment Plants (ETP) in their premises and to discharge the effluents only after treating them to the TNPCB standards. For less affluent tanners, who could not set up ETPs, either due to high cost or due to lack of space in their tanneries, the Tamilnadu Leather Development Corporation (TALCO) came forward to put up CETPs in Vaniyambadi and Ranipet. Nearly 55 tanneries are connected to the CET plant to treatment.

River Cauvery

Cauvery the lifeline of Tamilnadu is the eight largest river in the country originating 1355 meter above the mean sea level in the Western Ghats at Karnataka. With a running course of 765 Kms the river joins at Poombukar in Tamilnadu. The river is polluted when it passes through major urban areas. Industrial establishments at Mettur, Pallipalayam, Erode, Sirumugai and Trichirapalli discharge all types of effluents to the extend of 5082 lakh litres per day. Untreated sewage discharged into the river exceeds 80 lakh litre per day. In Salem district of Tamilnadu alone 20 million liter of effluent are discharged per day from Sago Mills.

REFERENCES

1. Sacred land film project –. http://www.sacredland.org/world_sites_pages/Ganges.html
2. Government of Tripura (2005), Tripura State Pollution Control Board "Pollution Status Assessment of Haora River, Tripura"
3. Centre for Science and Environment, New Delhi http://www.rainwaterharvesting.org/Crisis/River_Pollution.htm

4

Externalities of Waste Water Irrigation on Agricultural Land

Water is regarded as polluted, when it is discharged in the quality (or) composition, directly (or) indirectly as a result of human activities and it becomes useless (or) less suitable for drinking, domestic, agriculture, fisheries (or) other purposes for which, it would otherwise be quite suitable in the natural (or) unpolluted state. It reduces the amount of pure fresh water that is available for such necessities.

The real menace of water pollution arises from sewage, industrial waste and a wide array of synthetic chemicals being discharged into the water sources like river, streams and lakes. The factors responsible for water pollution may be natural (or) artificial. The natural factor of example, drought causes water level go down in aquifers and salt water intrusion may take place. During rain, running water gathers salt and other material. However more than the natural factors, the man- made factors, mainly originating from Industrial activities is the major responsible for produces negative externalities in the surrounding area, many industries, it may be either large, medium (or) small scale industries discharge their effluents, which usually join rivers and lakes, which are the main sources of water supply and biologically degrade the flora and fauna present in the river. When the pollution load in river exceeds in assimilating capacity, the additional pollution load adversely affects the health of the river and ultimately leads to further reduction in assimilating capacity. Finally a state is reached, when the content of the river, becomes toxic, unfit for any use.

Classification of Water Pollution

The water pollution is classified into five categories:

a) Organic Pollution

i) Oxygen demand waste included domestic waste, sewage, industrial waste and agricultural run-off. Most of them are bio-degradable organic compounds which are creating rotten and foul smell.

ii) Disease causing waste in sewage and other wastes would create pathogenic micro organism.

iii) Synthetic organic compounds which include waste materials arising out of man made synthetic materials, such as pesticides, insecticides, detergents, synthetic fibres, plastics, paints and industrial chemicals etc., which may enter the water course either by discharge on spillage during transport.

iv) Sewage and other agricultural run-off arises generally from agricultural land contains plant nutrients. This may stimulate growth of plenty of large and similar aquatic needs in the receiving water bodies.

B) Inorganic Pollutants

Finely divided metals, metal compounds, cyanides, sulphates, nitrates, mineral acids, inorganic salts etc, are the inorganic pollutants in water. A great variety of metal forms can exist in natural water at low concentrations. These metals and metallic compounds released from anthropogenic activities add up to their natural background levels in water. Some of these trace metals plays essential roles in biological process but at higher concentration they prove toxic.

c) Suspended Solids and Sediments are mainly due to soil erosion, sediments are mostly contributed by natural process of erosion, agricultural development, mining and construction activities. Suspended solids in water mainly comprise of silt, sand and minerals eroded from the load.

d) Radioactive materials water pollutants may arise from mining processing of ores e.g. Uranium tailings, use of radio active isotopes in agriculture, industrial and medical research and applicants radio active materials due to testing and also use of weaponry. They are toxic to any form of life if they exceed tolerance limit.

e) Health (or) heated effluents: This is the problem relating to discharge of heated effluents into the water bodies by industrial establishments, many industries using water as a coolant, dispose off the waste, and hot water by returning it into the organic water bodies.

Sources of Water Pollution

a) Domestic Waste

Domestic Waste is a very serious pollutant of wells

and river, which are important sources of drinking water. These river and wells are polluted by the human excreta besides that animals and birds. The drinking water from these sources contain high amount of *(1)* Nitrate, *(2)* BOD, *(3)* COD, *(4)* Chloride and Sulphite and *(5)* TDS. These effluents have high concentration of toxic and destroy fish and plant life.

b) Industrial Waste

The solid and liquid wastes of Industries play a significant role in polluting water. The industries such as *(1)* Pulp and paper, *(2)* distillery, *(3)* Pottasic fertilizer, *(4)* Electro plating plant, *(5)* Asbestos, *(6)* Alcohol, *(7)* Textile units, *(8)* Tanneries, *(9)* Sugar manufacturing, and *(10)* oils etc. are the major polluting industries.

There is an increasing trend to require more efficient use of water resources, both in urban and in rural as it uses more for agricultural purposes. In developing countries, fast growing population demands fresh water for domestic and agricultural use. But these countries have the lack of the comprehensive waste water management and pollution control technologies a major portion of the waste water pollute the natural water bodies. These polluted sources are used in villages and cities for agriculture and other purposes. In dry season, farmers often use the waste water from drainages and sewages because it is the only reliable source of water. Generally these waters are discharged into lagoons (or) dumped on low lying areas without any pre treatment thereby creating sewage pools, contaminating ground waters and salinizing good quality lands. At many places this waste water is discharged into drains and rivers causing serious water pollution.

In rural and peri-urban areas of most developing countries, the use of waste water for irrigation is a fact of life, not a matter of choice. In some cases, irrigation canals (or) rivers carry domestic wastes from upstream villages and down. In semi urban areas, the local sewage is the only water sources that support the livelihoods of millions of poor peoples who irrigate high value crops. Recent surveys across 50 cities in Asia, Africa and Latin America show that waste water irrigation is a common reality in three fourths of cities. In Vietnam and Pakistan alone, between 10,000 and 30,000 hectares of land are cultivated with undiluted waste water. This does not reflect large areas using diluted waste water (or) Polluted water. In Ghana, in and around the city of Kumasi alone, farmers use polluted water sources on about 12,000 hectares more than twice the area covered by the country's formal irrigation schemes[1].

The Mezquital valley in Mexico, which is probably the largest irrigation area, using raw waste water, covers more than 90,000 hectares. The country alone accounts in total for about half of the 5,00,000 hectares irrigated with waste water in Latin America. According to a survey, Global estimates of the total area under raw and diluted waste water irrigation are still fragmentary, but it might range from around 3 to 3.5 million hectares, with the largest share probably in China. In Australia and the USA waste water is used in agriculture, releasing high quality water supplies for potable use. Some countries e.g. Hashemite Kingdom of Jordan and the Kingdom of Saudi Arabia, have a national policy to reuse all treated waste water effluents and have already made considerable progress towards this end[2].

There is an increasing trend for more efficient use of water resources for domestic and agricultural purposes. The reuse of water for agricultural irrigation is often viewed as a positive means of recycling water so that the potential large volumes of water can be used. Recycled and well treated effluent water that can have the advantage of being a constant, reliable water source and reduce the amount of water extraction from the ground. On the other hand there are concerns about the impact of the quality of the recycled water both on the crop itself and on the end users of the crop. Water quality issues that can create real (or) perceived problems in agriculture include nutrient and sodium concentrations, heavy metals and the presence of contamination such as human and animal pathogens and pharmaceutics. Social attitudes to the use of crops that have been irrigated with recycled waters and the resulting impact on market value of crops are also major considerations. Hence the use of the effluent/waste water creates both positives and negatives effects on the agricultural economy.

Table 4.1: Effects of Waste Water on the Agricultural Economy

Positive Effect	*Negative Effect*
1) Conserves Water.	1) Health risk for irrigators and communities with untreated effluent users
2) Low cost method for sanitary disposal of municipal waste water.	
3) Reduces pollution of rivers, canals and other surface water resources	2) Contamination of ground water
	3) Build up of chemical pollutants in the soil (Heavy Metals)
4) Conserves nutrients, reducing the need for artificial fertilizer	
5) Increase crop yields	4) Creation of habitats for

6) Provides a reliable water supply to farmers.	disease vectors
	5) Entrophication of canals carrying waste waters.

Positive Effects of Waste Water Irrigation

Domestic waste water has always been a low cost option for farmers to go for irrigated agriculture in water scarce regions because of its resources and the high nutrient content helps the farmers to fertilize their crops without spending substantial amounts on additional fertilizers. In addition to temporal and spatial water scarcity, along with the rising demand for water from industrialisation, growing population forced the farmers to go for waste water irrigation. The irrigation with treated/untreated industrial effluent is a new practice since it is: *(i)* a low cost option for waste water disposal *(ii)* as a reliable assured and cheap source for irrigated agriculture, especially in water-starved and semi-arid areas *(iii)* as a way of keeping surface water bodies less polluted *(iv)* as an economic resource for agriculture due to its nutrient value.

Whenever good quality water is scarce, water of marginal quality will have to be considered for agriculture. The marginal quality water can be defined as water that possesses certain characteristics which have the potential to cause problems when it is used for an intended purpose. For example brackish water is a marginal quality water for agricultural use because of its high dissolved salt content, and municipal waste water is also a marginal quality water because of the associated health hazardous. From the view point of irrigation it has a significant impact on the quality of land.

Properly planned use of municipal waste water and well treated water by the industry alleviates surface water pollution problems and not only conserves valuable water resources but also takes advantages of the nutrients contained in sewage to grow crops. The availability of this additional water near the villages will increase the choice of crops which farmers can grow. The nitrogen and phosphorous content of sewage might reduce (or) eliminate the requirements for commercial fertilizers. It is advantageous to consider effluents reuse at the same time as waste water collection treatment and disposal are planned, so that sewage system design can be optimized in terms of effluent transport and treatment methods. The cost of transmission of effluent from inappropriately sited sewage plants to distant agricultural land is usually prohibitive. Additionally, sewage treatment techniques for effluent discharge to surface water may always be appropriate for agricultural use of the effluent.

The use of waste water can produce Agronomic and economic benefits of irrigation. Waste water can often contain significant concentration of organic and inorganic nutrients, for example nitrogen and phosphate. There is potential for these nutrients present in recycled water to be used as a fertilizer source when the water is recycled as irrigation for agriculture. To an illustration, a city with a population of 5,00,000 and water consumption of 200 ltr per person would produce approximately 85,000 m^3/d of waste water, assuming 85 percent is flow to the public sewage system, if treated waste water effluent is used in carefully controlled irrigation at an application rate of 5000 m^3/Ha in a year, an area of some 6000 ha could be irrigated. In addition to the economic benefit of the water, the fertilizer value of the effluent is of importance with typical concentrations of nutrients is treated waste water effluent from conventional sewage treatment process as follows.

a) Nitrogen (N) – 50mg/1 b) Phosphorous (p) - 10mg/1c c) Potassium k-30mg/1 and assuming an application rate of 5000 m3/ha.year, the fertilizer contribution of the effluent would be N – 250kg / ha.year, p – 50kg / ha.year, k-150kg / ha. Year

Thus all of the Nitrogen and much of the Phosphorous and Potassium, normally required for agricultural crop production would be supplied by the effluents.

The reuse of water is just one source of water that has potential for in agricultural settings. Reused water does have a major advantage in that it is usually a constant and reliable supply, particularly with sources such a treated with sewage or industrial effluent discharge. As well as being a constant source of water, much water suitable for reuse are produced in large volumes, which if not used would be merely discharged into the environment. The uses of waste water not only increase the level irrigation but also provide job and income for the traders who market the produce of the lands, input suppliers and other service providers.

Negative Effects of Effluents Water

Even though the use of waste water can produce some positive effects on the agricultural economy, there are number of risk factors associated for using reused water for purposes such as agricultural irrigation. Some rise factors are short term and vary in severity depending on the potential, animal (or) environmental contact (e.g.

Microbial pathogens) while others have longer term impacts which increase with continued use of recycled water. (e.g. saline effects on soil)

The waste water irrigation poses several threads to the environment via contamination by nutrients heavy metals and salts. The major disadvantages of waste water use are as follows.

a) Salinity

Salinity problems occur when the total quantity of soluble salts in the grass root zone is high enough to adversely affect the crop. Most effluent water are high in salts and they might accumulate the levels intolerable. The salinity of recycled water can impact both on the soil itself, as well as influencing the growth of the crops being irrigated. Salinity is the form of sodium can directly affect soil properties through the phenomena of swelling and dispersion[3].

b) Toxic Elements

Effluent water usually contains a wide variety of elements in small concentrations. Problems can occur when certain elements accumulate in the soil to levels toxic to plants. Toxicities can occur due to an accumulation of boron, chloride, copper, nickel, zinc or Cadmium[4].

According to Sheikh[5] (1987) heavy metals are easily and efficiently removed during common treatment process and the majority of the heavy metals concentrations in raw sewage end up in the bio solid fraction of the treatment process with very low heavy metal concentration present in the treated effluents. Thus heavy metals are of little concern of irrigation of crops when using treated effluents as a source of recycled water. If the source for the recycled water is from an industrial source or is less treated than normal then the influence of heavy metals would need to be considered. Heavy metals that are present in effluents used for irrigation tend to accumulate in the soils where there is a potential that they could become bio available crops.

Angelova[6] (2004) observed that fibre crops such as Flax and cotton did take up heavy metals when grows in heavily contaminated soils however the concentrations detected in the leaves and seeds was only a small percentage of the concentration present in the soil. A study was conducted by Ofusu[7] (1999) found that the crops irrigated with industrial waste water posed to health risks especially in heavy salts.

Sodium and other forms of salinity are the most persistent in recycled water and are among the most difficult to remove from water usually requiring the use of expensive caution as well as influencing the growth of the crops being irrigated exchange resins. The salinity of recycled water can impact both on the soil itself. Salinity in the form of sodium can directly affect soil properties through the phenomenon of swelling and dispersion. One of the major impacts on the soil is the reduction of hydraulic conductivity. This can impact on the ability of water to infiltrate into the soil profile and thus, reduce the water availability to irrigated crops. The salination of soil through the application of irrigation water make high salinity and the subsequent effect on clay in the soil is one mechanism that reduces hydraulic conductivity.

The quality of the reused effluent soil characteristics and initial quality of the receiving ground water all are important factors in determining the extent to which salt reaches and impacts the quality of the ground water. If the groundwater already is saline (or) has a high salt concentration then additional salt entering the ground water form the soil will have limited impact, where groundwater has lower salt concentrations. The impact of some additional salt may not be considered to adverse it. If the extent of ground water movement is limited, the ground water is not used for any other purpose (e.g.) drinking water supplies (or) the groundwater does not have discharge zone, where the salt can discharge into rivers and other surface water bodies. The other issues relating to elevated sodium levels in recycled waters, particularly effluents such as treated sewage is the effect the sodium has no crops irrigated with this water. Grain crops such as saline wheat is more resistant to the irrigation of soils with less drops in yields over a wide range of electrical conductivities compared with other more sensitive crops.

Health Consideration: The biological composition of effluent water is of great concern because of pathogenic bacteria and viruses. Effluent waters are not generally released for irrigation without prior approval of public health authorities. Some effluent water released for landscape irrigation is generally secondary effluent, it may contain some harmful chemical and biological substance. The irrigation posses a number of potential rises to human health via the consumption of (or) exposure to pathogenic microorganism heave metals, harmful organic chemicals such as endocrine disrupting compounds and pharmaceutically active compounds. Of these, pathogenic micro organisms are generally considered to pose greatest threat to human health. A wide variety of pathogenic micro organisms is found in waste water, including bacteria, viruses, protozoan, and parasitic worms. This symptoms and diseases associated with such infections are also diverse including typhoid, dysentery, diarrhea, vomiting and mal absorption. The concentration of pathogens in waste water is dependent on the source population and the susceptibility to infection carries from one population to another[8].

Waste Water Irrigation Problems in India

Waste water is gaining popularity as a source of irrigation water in different countries around the world. This is especially true in India, where it has been in use for a long time. Both food crops and cash crops are irrigated by domestic water and industrial effluent in India. The lack of alternative sources of water has generated markets for waste water. Throughout India Industries recycle water to reduce the requirement for fresh water. This trend led by industries in Saurashtra, Gujarat and Chennai, Tamilnadu[9]. Instances of industrial effluent disposal on land for irrigation are very limited in developed countries like the USA, UK, Canada and Australia. In India having the option to dispose effluents on land encourages the industries to discharge their effluents either on their land (or) on the surrounding farmlands in the hope that it will get assimilative in the environment through percolation, seepage and evaporation without any environmental hazards. Environmental problems related to industrial effluent disposal on the land have been reported from various parts of India and other countries[10]. In India 12 percent of people get clean drinking water while the rest 88 percent quench their thirst from polluted lakes, tanks, rivers and wells due to which more than three million people get affected or die of enteric diseases every year[11].

In a developing country like India, Industrial effluents as well as hospital and commercial waste often get mixed with domestic sewage, and unlike developed countries where Industrial effluents often get mixed with domestic sewage to dilute industrial pollutants and toxicants for better/easier treatment. In India mostly urban diffused industrial units disposed their untreated effluent in public sewers as a regular practice to avoid the cost of effluent treatment. In India only 24percent of waste water is treated before it is used in agriculture and disposed into rivers, and that is also for metro cities and class-1 cities[12]. The Indian pollution abatement rules and regulations provide options to industries to dispose their effluent in different environmental media e.g. on surface water bodies, on land for irrigation in public sewages (or) marine disposal, according to their location, convenience and feasibility. In the report of Central Pollution Control Board of India (2001) there are different prescribed standards for different disposal options. As far as Industries are concerned, their objective is to meet any one of those standards, which is feasible and convenient for them to discharge their effluent. The standards are set with the assumption that the environmental media have the capacity to assimilate the pollution load so that no environmental problems will rise.

Table 4.2: Maximum Possible Limit for Industrial Effluent Discharge

Parameter	*Into inland surface waters Indian Standards 2490 (1974)*	*Into public sewers Indian Standards 3307 (1974)*	*Onland for irrigation Indian standards 3307 (1974)*
pH	5.9 - 9	5.5 – 9.0	5.5 – 9.0
Biological O_2 demand (for 5 days at 20°C)	30	350	100
Chemical O_2 demand	250	-	-
Suspended solids	100	600	200
Total dissolved solids (inorganic)	2100	2100	2100
Temperature (°C)	40	45	-
Oil & grease	10	20	10
Phenolic Compounds	1	5	-
Cyanides	0.2	2	0.2
Sulphides	2	-	-
Flourides	2	15	-
Total residual Chlorine	1	-	-
Pesticides	-	-	--
Arsenic	0.2	0.2	0.2
Cadmium	2	1	-
Chromium (hexavalent)	0.1	2	-
Copper	3	3	-
Lead	0.1	1	-
Mercury	0.01	0.01	-
Nickel	3	3	-

Selenium	0.05	0.05	-
Zinc	5	15	-
Chlorides	1000	1000	600
Boron	2	2	2
Sulphates	1000	1000	1000
Sodium(%)	-	60	60
Ammoniacal Nitrogen	50	50	-
Radioactive materials			
Alpha emitters (millicurie/ milliliter)	10^{-7}	10^{-7}	10^{-6}
Beta emitters (μ curie/ milliliter)	10^{-6}	10^{-6}	10^{-7}

Source: TERI Energy Data Directory Year Book, 2000-2001.

Water Pollution by Industries

By assuming the concentration of pollution and the water consumption the industries are classified into 3 major groups depending on their general characters.

I. Major Water Intensive Industries

1. Tanneries and slaughter houses, 2. Pulp and paper Industries, 3. Dyeing, printing and processing of textiles, 4. Breweries and soft drinks, 5. Dairy and dairy Products, 6. Ice and cold storage, 7. Small steel mills, and 8. Food processing and preservation.

II. Moderate Water Consuming Industries

1. Canning Industries, 2. Roller and Flour mills, 3. Food Products like Biscuits.

III. Industries Using Water for Sanitation

1. Electrical and Electronic industries, *2.* woods, *3.* Rubber, *4.* Wax, *5.* Glass, *6.* Printing Press, *7.* Readymade Garments, *8.* Plastic, *9.* Paint and Varnish, *10.* Metal and Alloys, *11.* Textile and spinning, and *12.* Cement and ceramic products.

Table 4.3: Concentration of Pollutants in Major Industries in India

Sl No	*Name of the Industries*	*BOD Mg/l*	*SS mg/l*
1	Chemical Industry	1200	900
2	Water Intensive Industry	500	700
3	Engineering Industries	150	250
4	Industries using water for Sanitation	200	300

Source: Kudesia- Water Pollution, Pragat Prakkash Publishers, 2003.

Water Pollution by Small Scale Industries

Most of the environmental defaulting industries in India are major industries such as Sugar, distilleries, leather processing and thermal power stations. But they are capable for treating industrial effluents, but this is not the case with small scale industries, which can't afford enormous investments in pollution control equipments with appropriate licensing formalities, as their profit margin is slender. Today, the SSI's are getting adopted for and wide western technological systems, which produce enormous gaseous liquid and solid wastes. However, in under developed countries like India numerous Small Scale Industries (SSI) are too small to avail of the pollution control technology developed by the advanced countries. These enterprises come into existence because there is need for non-farm employment, particularly in countries where rural poverty and unemployment is high[13].

In India, the government has strong policies to promote small scale Industrial sector since independence for the reasons such as: *1)* They are relatively labour intensive and thus create more jobs, *2)* they contribute to decentralized industrial development, *3)* they are flexible and quickly reorient themselves to emerging demand, and *4)* export oriented.

The village and Small Scale Industries are categorized in three groups viz., traditional industries, house hold & cottage industries and modern industries. These enterprises work within an extremely competitive environment and restricted financial reserves to invest in non-productive and expensive pollution control technologies. Technical skills within these establishments are very low. They have a strong resistance to pollution control laws. The advanced countries have not developed pollution control technologies economically suitable for these enterprises. So these units often use outdated and highly polluting technologies and caused to generate pollution in the surroundings area.

SSI has become a major target of pollution control regulatory authority. These enterprises tend to form strong clusters with serious water pollution problems leading to utmost contaminating. In respect of waste water generation, about 40 percent of total industrial waste water are generated by small scale industries. It has been estimated that the total waste water generated out of these industries amounts to 3900 million litres per day (or) 1.2Gm^3 per year[14]. On the other hand the large scale industries, for instance the Tanning industry produce over 148 million cubic litres of water every year. Since the latter part of the 1980's tanneries have emerged as a major source of pollution. Their effort to duck the problem has led to a spate of judicial orders for the establishments of common effluent treatment plants.

But the establishments of common effluent Treatment plant remain elusive for SSI because of higher fixed cost. As stated by Maria[15] with the adoption of the water act, those small size industries had in theory the obligation to treat their effluent in order to reach a pollution concentration respecting the minimum acceptable standards laid down by the Pollution Control Boards (PCB). Nevertheless the size of these facilities makes the installation of a standard Effluent Treatment Plant (ETP) unaffordable because of the important fixed cost of an individual ETP. Therefore public authorities have taken the initiative to promote ETP scheme, allowing small industries to gather in order to treat jointly their effluents. Although CETP's are mainly seen as a mean to take advantage of scale economies, these schemes also act as a subsidies from public powers to small industries in order to allow them to respect the standards. The minimum participation asked from SSI in the CETP schemes implemented in India is 20 percent. The rest is funded through subsidies from central and state governments as well as loan from international organisation.

Table 4.4: Water Pollution by SSI

Sl.No.	*Industry*	*Cluster Region/ Products*	*Waste water generated (million l/day)*
1	Engineering	Hand Tools-Jalandar & Nagpur; Locks-Aligarh Bicycle parts - Ludhiana	2125

		Brass parts – Jamnagar Diesel Engine Parts-Agra, Kholapur, Rajkot, Ghaziabad	
2	Paper and Board Mills	-	1087
3	Textile Industries	Cotton and Knitwear, Hosiery- Tiruppur, Ludhiana, Calcutta, Delhi	450
4	Tanneries	Leather products- Chennai, Calcutta, Agra and Kanpur	50
5	Organic Chemicals	-	60
6	Pharmaceuticals	-	40
7	Soaps & detergents	-	10
8	Paint & Varnishes	-	10
9	Petro Chemicals	-	10
10	Edible and Vanaspathi	-	7
11	Dye & Dye intermedians	-	32

Source: Anil Agarwal, Centre for Science and Environment, New Delhi.

In early periods the pollution levels of SSI was low, since the effluent load would be more or less within the assimilative capacity of the surrounding environment. But the gradual increase in the effluent load in proportion with increased output had begun to serious environmental impacts. In addition to this the industries provide regional socio-economic benefits in the form of income, employment and foreign exchange. Since the SSI have historically, been promoted as a matter of industrial policy, the government agencies face a dilemma, in regulating the acute pollution problems caused by the SSI sector.

Externalities of Water Pollution—A Theortical Approach

Environmental pollution is an externality in welfare Economics. An externality is present whenever individual A's utility and production relationships includes real variables, whose values are chosen by others without particular attention to the effects of A's welfare. An externality can be either beneficial (positive) (or) harmful (negative). A neighbour's rose garden entering as a real variable (non-monetary) in the utility functions of others living in the neighborhood, similarly the services of a lighthouse entering as a real variable in the production function of shipping companies are the case of beneficial externalities. All kinds of pollution impact on others include negative externalities viz., aircrafts using an airport, which enters as a real variable in the utility functions of persons living in the neighborhood.

The pollution of a river which enters as a real variable in the production function of water supply undertakings drawing from the river of agriculture. This is explained that, industries (or) towns withdraw large quantities of waters for their use and / (or) discharge almost an equivalent of waste water they cause an externality problem to these farmers and others. Their actions have an economic impact on other users, viz. health problems and productivity in the basin. Any pollution sheltering activities (or) avoidance of pollution abatement costs in terms of disposal of untreated, partially treated (or) diluted industrial effluents on land (or) surface water bodies could transfer a large cost to society in terms of environmental pollution and related human health problems.

The case of a negative production from river water pollution is shown in figure (1). The market assumed to be competitive. The goal of profit maximization does perfect competitive firm in equilibrium to produce at a level of output at which price equals MC. The profit maximizing output decisions of the firm is the produce output at Q1, where the Marginal Private Cost (MPC) equates the Marginal Revenue (MR). The MPC includes the incremental cost of labour and capital to the firm. A firm's private output decisions take no account, however, of the external cost of production it imposed on the society via., the pollution that the firm created as a by-product. The external cost as represented by the Marginal External Cost (MEC), they are due to: (i) Loss due to salinisation of the land. (ii) Loss due to fall in employment in agriculture. (iii) Loss of agricultural production increased as output of the firm increased. The cost of the society (i.e. Farming community and local residents) is represented by the Marginal Social Cost (MSC) curve. Thus the welfare maximizing output from the point of view of society is Q′, where MSC equates MR.

The validity of the result once the externality of production is introduced. The implicit assumption is that all given cost are borne by the firm and thus may be termed as private costs. The market i.e., the polluted firms would produce an output, which is higher than the optimal output from Society's point of view (Q1 to Q) that is those marginal units of output from Q to Q1. The cost of the society of producing them

Figure No.1

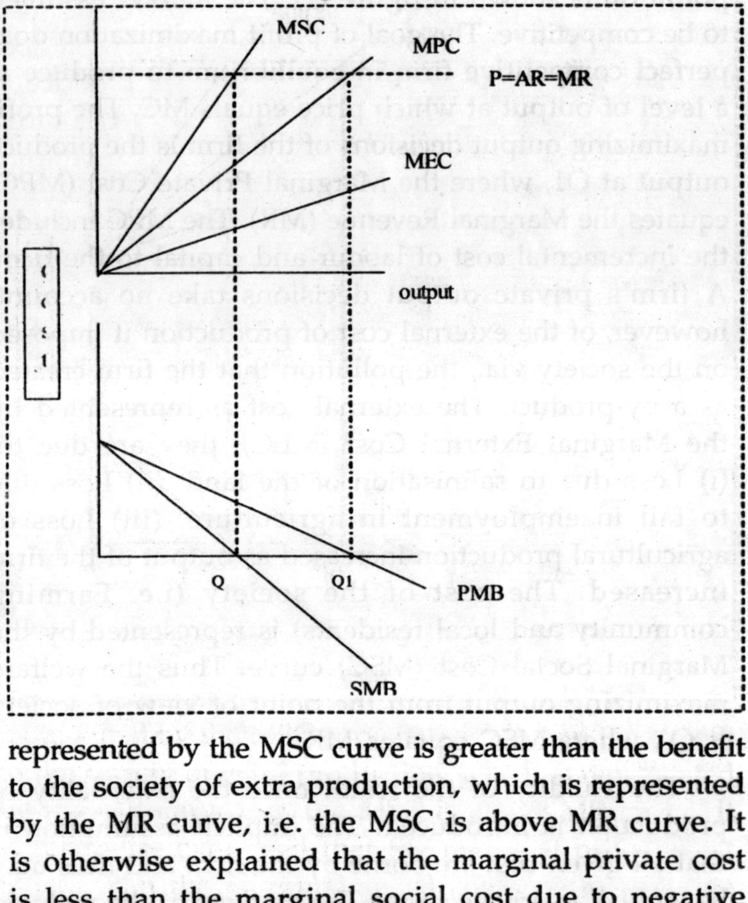

represented by the MSC curve is greater than the benefit to the society of extra production, which is represented by the MR curve, i.e. the MSC is above MR curve. It is otherwise explained that the marginal private cost is less than the marginal social cost due to negative diseconomies, since the firm will not take into account the disadvantage caused to the society by its activity.

Further the figure shows that a downward sloping Private Marginal Benefit (PMB) (or) utility curve which represented the private producer surplus accruing to the firm and is derived from the area below the price line above the firm's MC. Similarly the output decisions of the firm to produce output Q1 is one at which total net private benefit maximized best beyond the output of the private marginal benefit was negative. However, again the society faces the external cost of pollution from production and MEC curve, the efficient level of output from the society is again Q, where the PMB equated with MEC Beyond this output, the cost of the society of additional unit production is higher, since the factories will not take into account the external dis economies of scale. Correspondingly we can draw the Marginal Social Benefit (SMB) curve in a analogous way to the MPB curve, using MSC curve. From the SMB curve we could see that the society would produce output where PMB becomes zero, as a result when external diseconomies present, the output determined on the basis of private marginal benefit will be more than the socially optimum level.

REFERENCES

1. Simon Toze (2004): "Reuse of effluent water benefits and Risks" Paper presented at International Crop Science Congress, September, Brisbane, Australia.

2. *Ibid.*

3. Halliwal, D.J., Barlow, K.M. & Nash, D.M. (2002) "A review of the effects of waste water sodium in soil properties and their implications for irrigation system", *Australian Journal of Soil Research*, vol. 39, pp. 1259-1267.

4. Ali, Hariwandi (1982) "The use of effluent water for Turf grass irrigation", *Cooperative Extension*, vol. 32, No. 3 & 4.

5. Sheikh, B., Jaques, R.S. and Cort, R.P. (1987) "Reuse of tertiary municipal wastewater effluent for irrigation of raw-eaten food crops: a five year study". *Desalination*, Vol 67 Pp245-254

6. Angelora,V., Janov, R., Delibaltora, V. (2004) "Bio-accumulation and distribution of heavy metals in fibre crops" *Industrial Crops and Products*, Vol. 19, pp. 197-205.

7. Ofuso N. Asiedu. K. Otieno N.A, Omolo Ethengi (1999) Sewage reuse for irrigation in Athi river Town Kenya : Its

implications on public health, *Water Science Technology*, Vol. 33, pp. 343-346.

8. *Op.cit*, Simon Toze.
9. Bhamoria (2004) " Waste water irrigation in Vadodara, Gujarat: Economic Catalyst for marginalized communities" (ed), Christopher Scott, Fariqui and Liqa Rashid *Waste Water use in irrigated Agriculture—Confirming the livelihood and Environment Realties*, The International Development Research Centre, Ottowa.
10. *Op.cit.*, Prakash Nelliyat
11. Kudesia (2003)- "Water pollution", Pragat Prakkash Publishers, New Delhi,
12. Minhas P.S and Samra J.S (2004) " waste water use in Peri-urban agriculture- impact and opportunities", Central Soil Salinity Research Institute, Karnal, India.
13. Jagadees Pandi, S. (2008) *Incidence of Sickness in Small Scale Units*, Discovery Publishers, New Delhi.
14. Anil Agarwal " Small Scale Industries drives Indias Economy- But Pollutes Heavily" available at www.siwi.org
15. Maria, A. (2003) " The cost of water pollution in India", Paper presented at Conference on market development on waste and waste technologies through environmental economics' New Delhi, 30-31 October.

5

Impact of Effluent Water of Noyyal on Agricultural Land

Agriculture provides basic sustenance to all the living beings. It is very important that ecologically, socially and economically sustainable agriculture should become the backbone of the development process of an economy. Agriculture should be sustainable so that the natural resources such as soil, water and bio-diversity are used efficiently and equitably. It should be economically viable and lead to increasing employment opportunity, socially feasible, strengthening the role of women and other marginalized sections of the people

The rural economy is vastly agriculture oriented, historically has had policies in various phases for the development of agriculture with the expectation that development of agriculture would lead to overall development of the nation and help to minimize the unemployment and eradicating poverty. It has been recognized that the contaminated polluted water not only degrades the land but also affects the rural livelihoods and causes to increase the social cost of the farmers. As the polluted water percolates into the soil, large scale ecological losses are reported in crop land further leads to soil erosion, soil alkalinity and salinity, micro nutrient deficiency, water logging and depletion and contamination of ground water. These factors limit future gains from the land and water resources. The specific crops grown and the cropping practices employed also determine the residuals generated by the erosion and run-off. Irrigation is considered as the principle means of water loss from the natural system and it leads to arid condition downstream and ground water depletion.

Apart from on site costs reflected in the loss of productivity of soil, the off site costs due to agriculture is reported to be quite significant. The off site cost are caused by soil sediments transported in the surface water from eroded agricultural land. These include, river to public health. With the above background the present study focuses on the nature and extends of damages caused by the polluted Noyyal river water on the river bed area.

In this section an attempt is made to classify the study area into two zones, as the level of impact is different from one place to another. The two zones namely: *(i)* Tiruppur Taluk (South Block) of Coimbatore District and *(ii)* Perundurai Taluk of Erode District. Each Zones cover with five revenue villages.

Names of Revenue Villages

Zone –I (Tiruppur Tk)

Andipalayam ii. Muthalipalayam iii. Muthanampalayam. iv. Veerapandi and v. Mangalam.

Zone II (Perundurai TK)

i. Kodumanal ii. Orathupalayam iii. Ellaigramam iv.Siviarpalayam and v. Ramalingapuram

Table 5.1: Agricultural Details in Study Area

Sl.No.	*Details of Agriculture*	*Zone –I Land (in Hectare)*	*Zone –II Land (in Hectare)*
1	Geographical Details		
i.	Total Geographical Area	7808.64	8342.56
ii.	Irrigated Land	379.87	267.34
iii.	Non irrigated Land	1717.50	2615.89
iii.	Net Sown Area	2034.46	1849.25
iv.	Dry Land	3664.81	3895.36
2	Major Crops		
i.	vegetables &Fruits	11.90	8.87
ii.	Cotton	80.95	46.80
iii.	Coconut	260.22	187.54
iv.	Maize	1590.59	2115.76
v.	Other cereals& Millets	130.41	128.54
3	To. No. of Farmers	616	654

Source: Data collected and compilied from the reports of Divisional Statistical offices and Revenue villages.

It was a clear indication from the table (4.1) that the maize is the major crop of cultivation in both the Zones. It accounts for 1591 & 2116 hectares respectively in the study area, followed by coconuts cultivation for 260 & 188 hectares. Moreover the non irrigated area (1717.50 & 2616 hectares) were higher than the irrigated area (380 & 266 hectares). It could be understood that the major crops were rain-fed crops and the dry land is higher than the net sown area.

Table 5.2 Distribution of Respondents on the Basis of No. of Family Members

Sl.No.	*No. of Family Members (No.s)*	*Zone -I*	*Zone - II*	*Total % & out of total respondents*
1	Less than 4	22 (29.33)	19 (25.33)	41 (27.33)
2	5 to 8	41 (54.66)	30 (40.00)	71 (47.34)
3	More than 8	12 (16.00)	26 (34.66)	38 (25.33)
4	Total	75	75	150 (100)

Figure in (%) Percentages.

It could be understood that 47.34 percent of respondents had the family members of 5 to 8, of which 55 percent belonged to zone I and 40 percent belonged to zone – II. Next to 27.33 percent of respondents had the family members of less than 4, of which 30 percent were from zone – I and 25 percent were form zone – II. It is followed by 25 percent of farmers had the family members of more than 8, of them 35 percent belonged to zone II and 16 percent were from zone I. The analysis of number of family members of the farmers becomes imperatives as it affects the major variables taken for the study.

Table 5.3: Distribution of Respondents on Year of Residence

Sl.No.	*Zones*	*Year of Residence*	*No. of Respondents*	*Total Respondents*
1	Zone - I	< 15 Years	Nil	(-)
		16 to 30 years	12	(16.00)
		> 30 years	63	(84.00)
2	Zone - II	< 15 years	2	(2.66)
		16 to 30 years	8	(10.66)
		> 30 years	65	(86.67)

Figure in (%) Percentages.

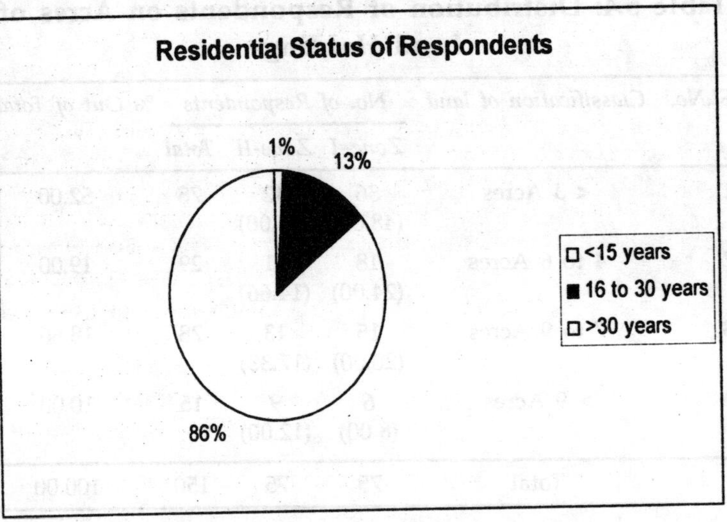

The above table indicates the year of residence of the respondents in the study area. It was identified that the farmers were residing for a long time and came to know that many of the farmers had a traditional living. The table (5.3) shows that 84 percent of the farmers in zone-I and 87percent of the farmers in zone-II were residing for more than 30 years. It is next to 16 percent of the farmers in zone-I and 11 percent of the farmers in zone-II were residing for 16 to 30 years. Only 2 percent of the respondent in the zone-II had been residing for less than 15 year. The analysis of the residential nature is imperative as it affects the variables of the study.

Table 5.4: Distribution of Respondents on Acres of Land Holding

Sl.No.	*Classification of land*	*No. of Respondents*			*% Out of Total*
		Zone–I	*Zone-II*	*Total*	
1	< 3 Acres	36 (48.00)	42 (56.00)	78	52.00
2	4 to 6 Acres	18 (24.00)	11 (14.66)	29	19.00
3	7 to 9 Acres	15 (20.00)	13 (17.33)	28	18.66
4	> 9 Acres	6 (8.00)	9 (12.00)	15	10.00
	Total	75	75	150	100.00

Figure in (%) Percentages.

As the marginal social cost of the farmers is different from one to another, because of the concentration of brackish water to the land, it becomes necessary to classify the farmers on the basis acres of holding. From the table (5.4) it was understood that majority of farmers were holding the (52 percent) land. less than 3 acres, of which 48 percent from zone I and 56 percent from zone II. These farmers were categorized as small and marginal farmers. Next to 19 percent of the farmers were holding upto 4 to 6 acres, and the same percentage of farmers were cultivating land upto 7 to 9 acres. These farmers were known as medium scale farmers. Only 10 percent of farmers were holding more than 9 acres of land. These farmers were categorised as large scale farmers.

Table 5.5: Distribution of Respondents on the Basis of Cultivation of Land

Sl.No.	Classification of years	No. of Respondents			% Out of Total
		Zone–I	Zone-II	Total	
1	> 10 Years	1(1.33)	3 (4.00)	4	5.33
2	11 to 20 Years	6(8.00)	5(6.67)	11	14.67
3	> 20 Years	68(90.66)	67(89.33)	135	90.00
	Total	75	75	150	100

Figure in (%) Percentages

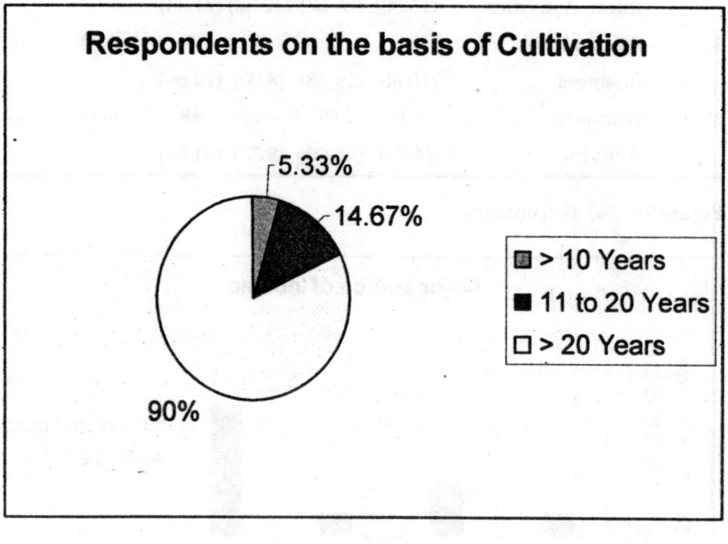

The table 5.5 had a corresponding nature with the previous table. As the farmers had a long residential nature, the year of cultivation too had a longer year. It was identified that nearly 90 percent of the farmers in both the Zone's had been cultivating their land for more than 20 years. It is next to only 15 percent of the farmers were cultivating land between 11 to 20 years, and only 5 percent of the farmers cultivating the land less than 10 years.

Table 5.6: Distribution of Respondents According to their Major Sources of Income

Sl.No.	Major Source of Income	Zone–I		Zone–II		% out of total respondents	
		Before the Impact	After the Impact	Before the Impact	After the Impact	Before the Impact	After the Impact
1	Agricultural farming	52 (69.33)	12 (16.00)	56 (74.66)	3 (4.00)	72	10
2	Allied Agricultural Activities	12 (16.00)	20 (26.67)	9 (12.00)	16 (21.33)	14	24
3	Agricultural Business	8 (10.66)	7 (19.33)	6 (8.00)	8 (10.66)	9.33	10
4	Non-farm Activities	3 (4.00)	36 (48.00)	4 (5.33)	48 (64.00)	9.33	56

Figure in (%) Percentages

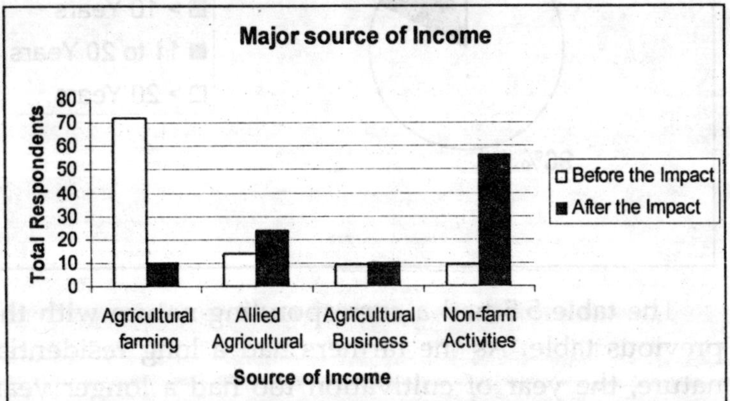

The percolation of effluent water not only affected the quality of land, but it also affected the source of livelihoods of the rural people. It is evident from the table (5.6) the source of income had become changed since the polluted water started to threaten their life.

It explained that only 10 percent of farmers were depending agriculture land to supplement their income, however previously it was 72 percent. Where as the non-farm activities increased from 9.33 percent in before the impact rose to 56 percent after the impact of waste water. It was understood that, as most of the farmers were having small size of land lead to severe economic pressures of their family and they find difficulties to get sufficient income to meet their basic necessities. The resources available cannot be spared to care for land, but indeed they may tend to migrate at any time.

The farmers supplemented their daily economic requirement form other than agricultural activities. By the penetration of the effluents, the activities of agriculture remain backward, and continued to be subsistence. The unresponsiveness and lack of the sector created stress and strain among the farmers by way of lower yielding per hectare, crop damage and their marginal social cost towards to the agricultural land have gone beyond the reach of middle and small farmers. The farmers in these areas are ready to getting out from farming as it becomes non viable.

Table 5.7: Distribution of Respondents According to the Sources of Water for Land

Sl.No.	*Sources of water*	*Zone – I*		*Zone – II*		
		Before the Impact	*After the Impact*	*Before the Impact*	*After the Impact*	*Overall % of Change*
1	River Water (37.33)	28 (14.37)	11 (18)	14 (12)	09	28% to 13%
2	Tank/Reservoir	NIL	NIL	38 (50.66)	NIL	50.67% to 0%
3	Tube well/ Open well	37 (47)	28 (32)	13 (17.33)	34 (45.33)	33.33% to 41.13%

4	Rain water	10 (13.33)	36 (48.00)	10 (13.33)	32 (42.66)	26.67% to 45.33%

The impact of pollution load in both the areas had significantly affected the sources of water for irrigation. It could be identified from the Table 5.7 that the people were using canal water from the river decreased to 13 percent at present from 28 percent before the impact. Whereas, the farmers using tube well and open well had increased to 41.13 percent from 33.33 percent. It was clearly understood that 45.33 percent of the farmers were relying only on rain water now, from 26.67 percent before the impact. While it was noted that none of the farmers were depending the sources of reservoir water. The reason can be traced that the farmers of zone-II had stopped to use the contaminated water of Orathupalayam reservoir. It is clearly evident that there was a shift in the sources of irrigation. The type of irrigation mainly practiced in both the zone was canal irrigation, in few areas large and medium level farmers were practicising drip irrigation.

According to the sources many peoples who had wells close to the Noyyal River and the dam site that the water had become bitter and salty and that there had been change in the water colour. It was chemical water and the wells had become naturally salty water. After the impact many farmers reported that they had drilled for a long feet the wells but they did not yield positive result. The sources of water for the small scale and marginal farmers were only rainwater and some part of lands of medium scale farmers also.

Table 5.8: Distribution of Respondents on the Basis of Crops of Cultivation

Sl.No.	*Major Source of Income*	*Zone–I*		*Zone–II*		*Overall % of change*	
		Before the Impact	*After the Impact*	*Before the Impact*	*After the Impact*	*Before the Impact*	*After the Impact*
1	Paddy	4 (5.5)	Nil	6 (2.00)	Nil	6.66	0
2	Cotton	11 (14.66)	2 (2.66)	6 (8.00)	5 (6.64)	10.66	4.66
3	Vegetables	13 (17.33)	6 (8.0)	10 (13.33)	2 (2.66)	15.33	5.33
4	Millets (maize)	30 (40)	58 (77.33)	31 (41.33)	63 (84)	62	80.67
5	Coconuts	16 (21.33)	7 (9.33)	16 (21.33)	3 (4)	21.33	6.67
6	Other (ground nut)	1 (1.33)	2 (2.66)	6 (8.00)	2 (2.66)	4.66	2.66

Figure in (%) Percentages.

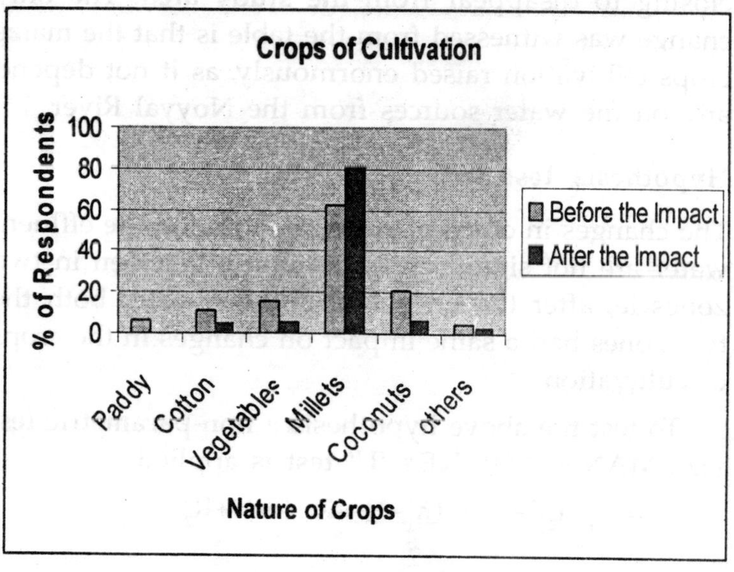

Before the impact of effluents the crops grown in the zone – I were cotton, coconuts, vegetables and Herbal products. In case of zone – II, before the construction of Orathupalayam dam, the crops grown were cotton, turmeric, onion, maize and coconuts. The coconuts, cotton and vegetables were cultivated in monsoon season while the maize crop was cultivated only in the season of summer, as it was a rain-fed crop. After the impact of the effluents the negative effects started to become obvious, and were noticed as inhibited germination and growth as well as a deterioration of the soil structure, where the rain-fed crop was able to withstand.

The table-5.8 indicated that the nature of changes in crop cultivation. It shows that the rice field in the total study area completely disappeared. Similarly the crops of cotton (from 10.66% to 4.66%), vegetables (15.3% to 5.33%) and coconut (21.33% to 6.77%) were closing to disappear from the study area. The only change was witnessed from the table is that the maize crops cultivation raised enormously, as it not depend any on the water sources from the Noyyal River.

Hypothesis Testing

The changes in crops of cultivation due to the effluent water are not significantly difference between in two zones ie, after the impact of effluent water both the two zones had a same impact on changes in the crops of cultivation.

To test the above hypothesis a non-parametric test viz., MANN-WHITNEY 'U' test is applied.

$$U = n_1 n_2 + \frac{n_1(n_1+1)}{2} - R_1 \text{ (or) } + R_2$$

The above method helps us to determine whether the two samples have come from identical populations. If it is true that the samples have come from the same populations, it is reasonable to assume that the means of ranks assigned to the values of two samples are more or less same.

U value	=	16.5
Mean of U	=	12.5
Varance of U	=	4.78
$Z = \frac{U - Uu}{\sigma}$	=	1.84

Since the calculated value of Z 1.84 is at 0.05 is 1.96 %, therefore we accept the null hypothesis at 5% level. It is concluded that the two studying zones had no difference in changing cultivation of crops by the intrusion of effluent water ie both of them had a similar changes.

Table 5.9: Distribution of Respondents According to the Impact Level of Effluent

S.No.	*Zones*	*Level of Impact/ Scale of farmers*	*No. of Farmers*			*Total & % out of Total*	
			Small >3	*Medium 4 to 9*	*Large Scale >9*		
1	Zone – I	i. Extreme	20 (26.66)	26 (34.66)	3 (4)	49	65.33
		ii. Moderate	9 (12)	4 (5.33)	1 (1.33)	14	18.66
		iii. Low	7 (9.33)	3 (4.00)	2 (2.66)	12	16.00
2	Zone – II	i. Extreme	34 (45.33)	19 (25.33)	7 (9.33)	60	80
		ii. Moderate	6 (8.00)	4 (5.33)	2 (2.66)	12	16
		iii. Low	2 (2.66)	1 (1.33)	Nil	3	4.00
	Total		78 (52.00)	57 (38.00)	15 (16.00)	150	100

Figure in () Percentages.

The study disclosed that the environmental impact of industrial effluents irrigation is different from one location to other as they have different assimilative capacities to absorb the pollutants. The Table 5.9 explained the impact of pollution level on farming lands. With regards to the land holdings the small scale farmers were seemed to be the highest victims of effluent water. Nearly 36 percent of them had fallen in the category of extreme level, followed by 30 percent of medium scale farmers and 6 percent of large scale farmers. In respect of the impact level 80perecentage of farmers in zone- II were affected by extreme level. It could be identified that the impact level of zone-II was relatively higher than the Zone-I. The main reason can be attributed that the sources of contaminated water emerged out from Orathupalayam reservoir.

The Orathupalayam dam was constructed at the cost of Rs.16.46 crores in 1991. The dam is situated 25km distance from Tiruppur and 12km distance from Kankeyam Taluk in Erode district, with the purpose of irrigation, to about 500 acres in Erode district, and 9875 acres in Karur District, Tamilnadu. The major source of water for this dam from Noyyal River. Instead of storing fresh and quality water, the dam received and stored the polluted Noyyal water. The problem of this dam was first noticed in the year 1993. Since then it started to contaminate all the water sources in and around 4km radius of the dam. From the focal discussions to the local residents, first the water started tasting bad. There was also bad odour. Later it spoils the irrigation source of bores and the wells. Even though water from the dam reaches the channel only for three to four months in a year, it was enough to contaminate the area. Now the dam becomes a mere storage tank for effluent of industries and the release of water had been stopped for irrigation purposes. Even though water has not been discharged from the dam at present, the effect that it had already contaminated the farming lands. Hence the farmers in the adjacent of this dam were extremely affected by the polluted water.

Hypothesis Testing

To obtain the variation within the subgroup and between the sub groups, Analysis of Variation (ANOVA) test is applied. The ANOVA is used to test whether the means of a number of populations are equal; the null hypothesis is stated as:

- There is no significance difference between rows i.e between the different scale of farmers.
- There is no significance difference in the level of impact of effluents and the different categories of scale of farming

Table 5.10: Calculated ANOVA Table for Impact Level of Effluent

Sources of variation	*Degrees of freedom*	*Sum of Square*	*Means of square*	*F-value*	$F_{0.05}$
Between the different scale of farmers (Between Rows)	2	666	333	3.0739	6.94
Between the level of impact and different scale of farmers (Between Columns)	2	1532.67	766.33	7.0740	6.94
Residual	4	433.33	108.33		
Total	8	2632			

For $V_1 = 2$ and $V_2 = 4$, $F_{0.05} = 6.94$, where the calculated value is 3.07 therefore, as the calculated value is less than the table value the hypothesis is accepted, which means that all the farmers had been affected by the effluent water.

For $V_1 = 2$ and $V_2 = 4$, $F_{0.05} = 6.94$. The calculated value of F (7.07) is higher than the table tabled value. In other words there is significance between the levels of impact of effluents and different scale of farmers. Hence the null hypothesis is rejected at 5% level and it was proved that the impact level were different among different scale of farmers.

Table 5.11

Yielding Per Hectare of Crops in Study Area

Sl.No.	*Crops*	*Zone – I Mean value*	*Zone – II Mean value*
1	Millets (maize)	1.68 Quintals	1.5 Quintals
2	Coconut	1000 nuts	700 nuts
3	Vegetables	4.5 Quintals	3.8 Quintals
4	Cotton	1Bales	1.2Bales

Increased agricultural productivity leads to increase in production, which in turn raise the income of the farmers. This increased income of the family community paves the way of augmenting savings, which can be used for further development of their agricultural land on the other hand a decrease in productivity leads to lowering the yielding per hectare of land and caused to suffer the farmers. As the effluents had been accumulating for the past 20 years and there is barely any agricultural land to be proved in terms of productivity.

The table 5.11 explains that the yielding per hectare

of land. It could be understood the mean value of maize were around 1.5 quintals per hectare, coconut were around 800 nuts, vegetables were 4 quintals and cotton had 2 bales per year. According to the farmers crop yielding per hectare drastically reduced. In these areas the same crops they could have yielded at an average of nearly 4 quintals of millet,s 8 quintals of vegetables, 5 bales of cotton and 10000 coconuts in no.s before the impact of pollution.

The soils in these areas are very exhausted because of the contaminated water. The land had lost all fertility (or) productive capacity and to revitalize fertility. The farmers in these areas are tradition bound and were not able to apply advanced techniques of production. More over the farmers complained that the seeds did not germinate any more when they continued to use the water for irrigation. According to them the effluents water was high in salts and they might have accumulated to the levels of intolerable to most of the crops.

Table 5.12

Distribution of Respondents According to Year of Flow of Effluent Water

Sl.No.	*Classification of year/Source of water*	*River water*	*Tank/ Reservoir*	*Tube/Open well*	*Rain water*	*Total & % out of total*
Zone I	a. less than 7 years	1	Nil	Nil	Nil	1 (1.33)
	b. 8 to 15 years	9	Nil	4	2	15 (20.00)
	c. More than 15 years	28	Nil	25	6	59 (78.66)

Zone II a. less than 7 years	Nil	2	Nil	1	3 (4.00)
b. 8 to 15 years	12	41	5	4	62 (82.66)
c. More than 15 years	10	Nil	Nil	Nil	10 (13.33)
Total	60 (40.00)	43 (28.67)	34 (22.67)	13 (8.66)	150

The year of flow of waste water differ from one zone to other. It could be understood that the farmers of Tiruppur area (zone - I) were relishing the problem since 1980's when the textile manufacturing units started to flourish. On the other hand the farmers in Perundurai Taluk (zone – II) were receiving the problem of effluent water at the midst of 1990's since the Orathupalayam dam started to contaminate the lands. Before the emergence of textile units at 1980's the water quality in all systems were good in Tiruppur areas, similarly before the establishment of the dam in Perundurai area. The data pertaining to the years of flow of water showed that 79 percent of the farmers in Zone-I and 18 percentage of farmers in Zone-II were receiving the problem of more than 15 years. But 83 percent of the respondents from Zone-II were affected for the years of 8 to 15. It showed that since the initial phases of industrial development in Tiruppur, only the down stream area were affected little, but it had started to affect the other river basins areas in the Erode and Karur districts. With regard to the source of water, the farmers who were using river water was highly affected (40 percent) next to Tank and reservoir (28.67), open wells (22.67 percent) and rain water (8.66 percent).

According to the farmers, in the initial stages the

pollution level from these units were low, since the effluent load would be more (or) less within the assimilative capacity of the surrounding environment. Hence the local public did not consider these bleaching and dyeing industries to be a problem. But the gradual increase in the effluent load in proportion with increased output has begun to have serious environmental impacts. Previously the water quality was comparatively good for irrigation and other uses. When the open wells were polluted the farmers had to depend on only rain water. However these sources of water had also concentration of pollutions since the effluents spread across the land. According to the farmers association in Zone-II, the long distance spreading of pollution to ground water was due to the stagnation of water in the Orathupalayam dam. Flowing water in the Noyyal River did not infiltrate to ground water to the same extend as when the river water is stored and stagnated for a long period of time. Because of the continuous discharge of effluent water and the damming of this water in the Orathupalayam dam the open and bore wells around the dam, which were almost free from pollution, have turned brackish. The rainfall over the area only marginally dilutes the pollution of well water but due to the high amounts of salt it can no longer be used for agricultural purposes.

Table 5.13
Distribution of Respondents by Sources of Waste Water

Sl.No.	*Zones/Sources of Waste water*	*Percolated with Ground water*	*Percolated with tank/ reservoir*	*Unstructured Sewage*	*Direct flow of effluent from river water*
1	Zone - I	21 (28.00)	Nil	8 (10.66)	46 (61.33)
2	Zone - II	26 (34.67)	35 (46.66)	Nil	14 (18.67)
	Total	47 (31.33)	35 (23.33)	8 (5.33)	60 (40.00)

N=150, Figure in () Percentages.

The table 5.13 analyses the source of waste water in the farming land in the study area. It was understood that, the effluent water directly flowed from river affected 61 percent of farmers in Zone-I and 18 percent at zone-II. They are next to 28 percent in zone-I and 35 percent in Zone-II were affected by the percolation of pollutants in ground water. It further explained that

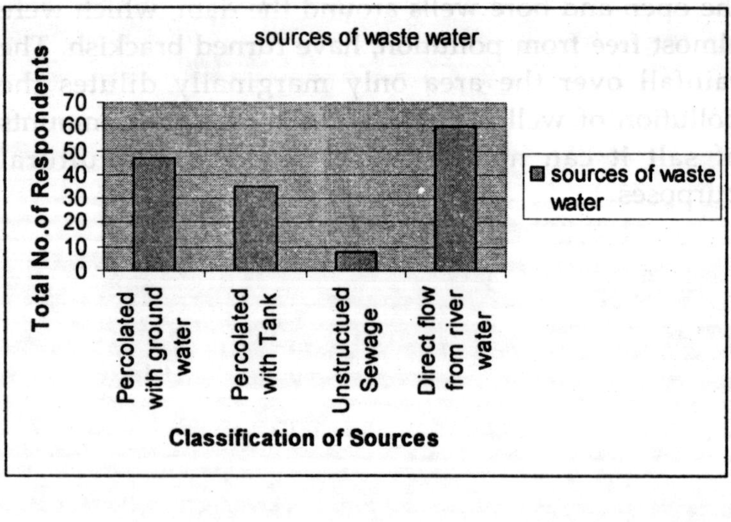

47 percent of farmers had the source of waste water from the percolation with toxins into tanks who were all the farmers in Zone-II and 10 percent of farmer in Zone-I were affected by unstructured sewages in Tiruppur area, as their lands were closer to the textile industries.

Table 5.14
Distribution of Respondents on the Basis of Conditions of Land

Sl. No.	*Zone*	*Acres of Land/ Condition of land*	*Most of them good*	*Half of them Good*	*Most of them poor*
1	Zone – I	a) < 3 Acres	Nil	6 (8.00)	30 (40)
		b) 4 to 6 Acres	Nil	7 (9.33)	11 (14.66)
		c) 7 to 9 Acres	2 (2.66)	4 (5.33)	9 (12.00)
		d) > 9 Acres	2 (2.66)	Nil	4 (5.33)
2	Zone – II	a) < 3 Acres	Nil	3 (4.00)	39 (52.00)
		b) 4 to 6 Acres	Nil	2 (2.66)	9 (12.00)
		c) 7 to 9 Acres	Nil	5 (6.66)	8 (10.66)
		d) > 9 Acres	Nil	3 (4.00)	6 (8.00)
	Total		4 (2.66)	30 (20.00)	116 (77.33)

N=150 Figure in () Percentages.

When environment degradation started it ultimately affects every one immediately and severe impact was felt by the poor farmer. The small farmers were exposed to have unhealthy farming practices. When land productivity declines, rich farmers were able to manage by increasing external inputs such as fertilizers and pesticides while the small and marginal farmers were fighting a losing battle and finally had lose themselves.

From the Table 5.14, the condition of land by the small and marginal farmers were mostly very poor, of which the farmers in Zone-II were very critical (83 percent), whereas the lands of large scale farmers was somewhat good in both the zones. However, all the large scale farmer indicated that their lands were not mostly good. It is understood that the effluent water not only polluted the ground water, but it has also damaged the fertile lands.

Hypothesis Testing

- There is no significant between rows, i.e acres of land possessed by the farmers.
- There is no significant difference between the condition of land and different scale of farming, i.e. the lands of all the farmers had poor conditions because of contaminated water.

To test the hypothesis ANOVA test is applied.

Table 5.15
Calculated ANOVA for Condition of Land and Scale of Farming

Sources of variation	*Degrees of freedom*	*Sum of Square*	*Means of square*	*F-value*	$F_{0.05}$
Between the rows (acres of land)	2	859	859	3.56	5.14
Between the different scale of farmers and the condition of land	3	769.667	256.67	1.0640	4.76
Residual	6	1447.333	241.222		
Total	11	3935			

For $V_1 = 2$ and $V_2 = 6$, F= 0.005 = 5.14. The calculated value of F (3.56) is less than the table vale the hypothesis is accepted at 5% level.

For $V_1 = 3$ and $V_2 = 6$, F0.05 = 4.76. The calculated value of F (1.06) is much less then the tabled value. There is no significant difference between the condition of land and different scale of farming and the above hypothesis is accepted at 5% level. Hence it is concluded that all the farmers had a poor condition of land for cultivation.

Table 5.16

Reason for the Deterioration of Land

Sl.No.	*Reason*	*Zone - I*	*Zone - II*
1	Loss of Nutrients in Land	75 (100)	75 (100)
2	Problem related to inputs of Agriculture	8 (10.66)	2 (2.67)
3	Soil degradation due to other reasons	5 (6.67)	Nil

Note: The Percentage of total exceeds above 100 as number of farmers cited more than one reasons.

The Noyyal river water is the root of all the sources of water in these areas whether be it an open or bore-well and dam. However, since the Noyyal water had turned the fertile soil as toxic with heavy metals, the salinity of this water could impact on soil itself as well as influencing the growth of crops being irrigated. This would have affected on the ability of water to infiltrate into the soil profile and thus reduce the good water availability for the irrigated crops. Invariably all the farmers reported that their land lost its nutrition level that was the chief causes of poor productivity level of agriculture. In addition to this, few farmers had added

Table 5.17

Opinion about Quality of Land

Sl.No.	*Quality of land/ Sources of Water*	*Zone I*				*Zone II*				*% out of Total*
		River water	*Tank*	*Tube open well*	*Rain water*	*River water*	*Tank*	*Tube open*	*Rain water*	
1	Deteriorated	24 (32.66)	Nil	30 (9.00)	10 (13.33)	14 (16.66)	38 (50.66)	18 (14.66)	8 (10.66)	135 (90)
2	No changes	4 (5.33)	Nil	7 (9.38)	Nil	Nil	Nil	2 (2.66)	2 (2.66)	15 (10)

N = 150 Figure in () Percentages

the problem related to input of agriculture (17.33 percent) and some other factors (2.67 percent) for the reasons for poor productivity of land.

With the virtual death of Noyyal, numbers of farmers have lost a vital water resource. Today the farmers and local fear to even touch the river water. In case, they venture out in to the water by mistake they develop boils and rashes out on their feet. The polluted water had deeply seeped underground, and thus contaminated all the water resources.

The present Table 5.17 compares the quality of land, irrigated by different sources of water. It explained that none of the farmers in all the water sources had positively inclined for the irrigation of effluent water. On the other hand 90 percent of the farmers felt that their land were deteriorated by the polluted water of which 40 percent belonged to the irrigation by tube well in Zone-I and 51 percent form the use of Orathupalayam tank in Zone-II. Only 10 percent of farmers expressed that their quality of land had no changes, who were all known as the cultivators away from (2 Kms) the Noyyal River bed.

From the available sources many plants were very sensitive to salt stress, especially during germination and the early phase of growth. Soil fertility is a factor that could affect how plants can tolerate a high concentration of salt. Ultimately it affects and deteriorates the quality of land. The effluents from the dyeing and bleaching industries contain part of the chemicals used for processing and treatment. Effluents vary widely in colour and are always forbid and highly alkaline. The various science studies showed that soil quality in these areas were alkaline.

Table 5.18
Distribution of Respondents by Nature of Changes in Crops Cultivation

Zones	*Nature of Changes/ Acres of Land*	*< 3 Acre*	*4 to 6 Acres*	*7 to 9 Acres*	*> 9 Acres*	*% out of Total*
Zone I	1) Intensive to Rainfed crops	30 (40.00)	14 (18.66)	13 (17.33)	4 (5.33)	61 (81.33)
	2) Rainfed to Intensive	Nil	Nil	Nil	Nil	Nil
	3) No Changes	6 (8.00)	4 (5.33)	2 (2.60)	2 (2.60)	14 (18.67)
Zone II	1) Intensive crop to rainfed crop	30 (40.00)	11 (14.66)	11 (14.66)	8 (10.60)	60 (80.00)
	2) Rainfed crop to Intensive crop	Nil	Nil	Nil	Nil	Nil
	3) No changes	12 (16.00)	0	2 (2.66)	1 (1.33)	15 (20.00)

Figure in () Percentages.

Crop diversification may be understood as a type of cropping system where any field crop does not contribute more than 50 percent of the total productivity and profitability. It is generally viewed as a shift from intensive cropping to rain-fed crops. These diversification of crop had also mounted the pressure on farmers as the rain-fed crops did not yielding high income and not most profitable. The Table 5.18 analyses the nature of crop diversification in the study area. It could be understood that majority of the farmers in all categories had a shift from intensive to rain-fed crops. An average of 80 percent of the farmers had changed the crop diversification from intensive to rain-fed. Only 18.6 percent of respondent in zone-I and 20 percent of respondent from Zone-II had not diversified their crops. But none of the farmers had indicated that they shifted from rain-fed crops to intensive crops, as the intensive crops require good water sources.

It is interpreted that in rain-fed areas productivity of crops is closed link with onset of distribution of rainfall. The small and marginal farmers, who are the most vulnerable community and to be backward in infrastructure, amenities and supporting services for agriculture, and particularly suffer from low investment on technology input. Intensified agriculture in rain-fed areas really requires better approaches, strategies and higher proportion of investment in future. But these poor farmers were not able to have the required investment for farming.

Table 5.19

Distribution of Respondents According to Effect of Effluent water

Sl.No.	Nature of Effect	Zone - I	Zone - II
1	Less of few crops	68 (90.66)	60 (80.00)
2	Amount decreased	69 (92.00)	72 (96.00)
3	Kept as Fallow land	41 (54.66)	66 (88.00)

* The total percent may exceed 100 as number of farmers, cited more than one reasons.

The majority of the places in the farming area of the two zones were suffered from the pollution load. The pieces of land do not give adequate work for most of the people depending on them. The farmers are seemingly engaged in agricultural activities and contribute very little (or) none to production. Because of the pollution load the farmers had to meet various problems from their land. The Table 5.19 explains that

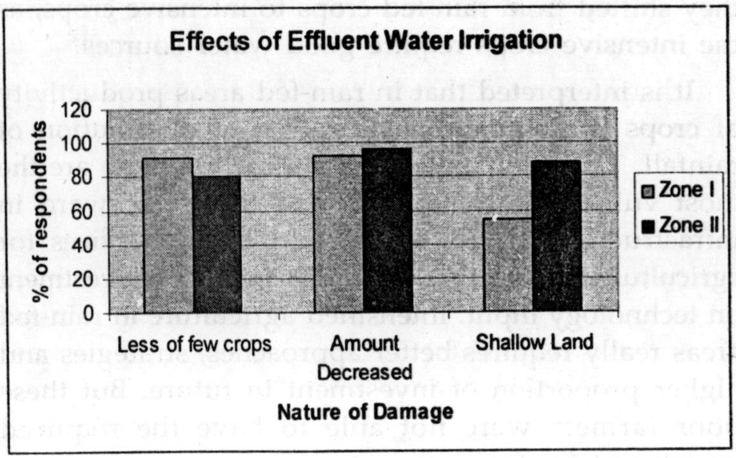

the farmers forced to yield low income by yielding less of few crops cultivation. It shows that 91 percent of respondents of Zone-I and 80 percent of respondents from zone II yielding less of few crops. Similarly at an average of 94 percent of farmers experienced with decrease in the amount from the activities of land. More over 55 percent of farmers in zone I and 88 percent farmers of Zone-II were keeping part of their land as fallow.

Since large numbers of farmers in the study area were small farmers, they cultivate small portion of land, and in many cases they work as casual labourer in the farming of large scale workers. With these small units of land they were not able to reap the benefit of economies of large scale of production. The pollution problem had further degraded their land as well as their livelihood. With little resources they could not make any big steps for raising the productivity of land.

The plight of farmers were aggravated once monsoon recedes as the river water leaves behind a

Table 5.20

Distribution of Respondents by Changes in Seasons of Cultivation

Sl.No.	Zone	Acres of Land/ Seasons of cultivation	Before the impact			After the impact		
			All the 3	2 Seasons	One Season	All the 3	2 Seasons	Only one
1	Zone - I	i < 3 Acres	18(24.00)	5(6.66)	13(17.33)	Nil	3(4.00)	33(44.00)
		ii. 3 to 6 Acres	10(13.35)	4(5.33)	4(5.33)	Nil	6(8.00)	12(16.00)
		iii. 7 to 9 Acres	12(16.00)	3(4.00)	Nil	6(8.00)	5(6.66)	4(5.33)
		iv.> 9Acres	5(6.66)	1(1.33)	Nil	2(2.66)	4(5.33)	Nil
2	Zone - II	i. < 3 Acres	24(32.00)	10(13.33)	8(10.68)	Nil	3(4.00)	42(56.00)
		ii. 3 to 6 Acres	10(13.00)	1(1.33)	Nil	Nil	3(4.00)	8(10.66)
		iii. 7 to 9 Acres	11(14.66)	2(2.66)	Nil	Nil	8(10.60)	5(6.67)
		iv. > 9 Acres	7(9.33)	2(2.66)	Nil	6(8.00)	3(4.00)	Nil
3	Total		97 (64.66)	28 (18.67)	25 (16.67)	14 (9.33)	32 (21.33)	104 (69.33)
	Zone - I		45 (60.00)	13 (17.33)	17 (22.67)	8 (10.66)	18 (24.00)	49 (65.33)
	Zone - II		52 (69.33)	15 (20.00)	8 (10.67)	6 (8.00)	14 (18.60)	55 (73.33)

Figure in () Percentages

blanket of organic matters on a large tract of agricultural land. These organic pollutants were deposited in the fields even in the dry season as the farmers use this polluted water to irrigate their fields. Large deposit of such slurry could be found in these places, which have adversely affected the productivity. Ultimately the helpless farmers forced to stop their cultivation.

The Table 5.20 explains that the changes in the season of cultivation of crops due to effluent water. The result of the study indicated that the farmers were able to cultivate at 3 seasons per year before the impact. After the penetration of the effluent water the seasons of cultivation had a drastic change. It showed that 69.33 percent of respondents in Zone-I and 60 percent respondents in zone- II were cultivating the land in all the 3 seasons, while 16.67 percent of respondents were cultivating only one season before the impact. At present only 9.33 percent of which 10.66 percent in zone-I and 8 percent in Zone-II were able to cultivate at 3

Table 5.21
Ranking the Major Reason for changing the Nature of crops

Sl.No.	*Major Reasons*	*Zone - I*	*Zone - II*	*% out of total respondents*
1	Problem related to water pollution	68 (50.67)	75 (100.00)	143 (95.33)
2	High Cost of cultivation	14 (5.33)	12 (16)	16 (10.67)
3	Urbanisation	18 (24.00)	Nil	18 (12)
4	Industrialization	5 (6.67)	Nil	5 (3.33)

Note: The percentage exceeds 100 percent, as number of respondents opted more than one reason.

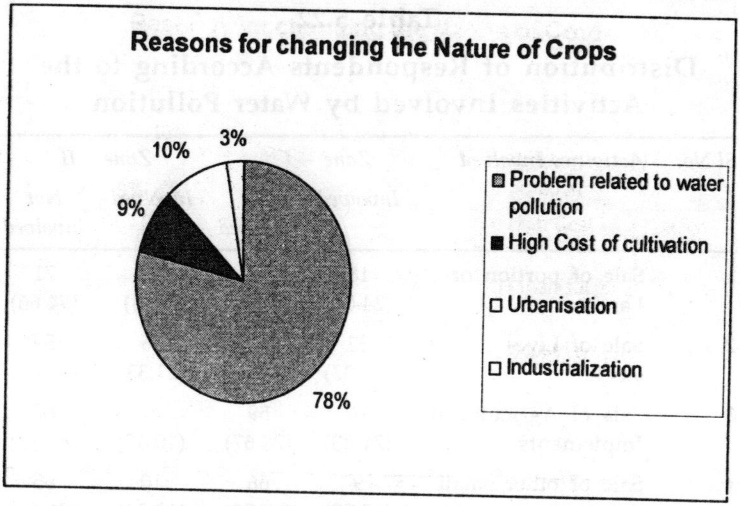

times. These farmers were known as large scale farmers and could manage by their large scale of production. On the other hand, 69.33 percent of the farmers cultivating their land for only one season and they were all small and marginal farmers.

The above table explains that the major reasons for changes in agricultural crops in the study area. It was identified that 95.33 percent of the respondents had changed the pattern of crops, i.e, intensive to rain-fed, because of the problem of water pollution. Few farmers reported that rise in prices of agricultural inputs, otherwise the high cost of cultivation (10.67 percent) was the another reasons. The farmers in zone I cited few other reasons such as urbanizations (12 percent) and industrialization (3.33 percent), as the area is closer to the city Tiruppur where industrial devclopment activities are carried out rapidly.

Table 5.22
Distribution of Respondents According to the Activities Involved by Water Pollution

Sl.No.	*Activities Involved*	*Zone – I*		*Zone - II*	
		Involved	*Not Involved*	*Involved*	*Not Involved*
1	Sale of portion of Land	18 (24.00)	57 (76.00)	4 (5.33)	71 (94.66)
2	Sale of Live-stock	22 (29.33)	53 (70.67)	16 (21.33)	59 (78.66)
3	Sale of Agricultural Implements	16 (21.33)	59 (78.67)	8 (10.67)	67 (89.33)
4	Sale of other small Assets	9 (12.00)	66 (88.00)	10 (13.33)	65 (86.69)

Figure in () Percentages.

The contaminated of polluted water is one of the main concern in the regions with limited water resources. Besides the fall in agricultural production and drinking water, their livelihoods were threatened. Every year after the post monsoon, the river leaves a thick deposit of pollutants on the banks, and starts to contaminate the agricultural land. The farmer uses the water of Noyyal has now paid a huge price for it. A portion of their land was rendered fallow as the problem continuous year after year, the land turns grey and the farmers were unable to grow any crops in the field. This in turn affects the farmers both directly and indirectly such as severe drought (or) continuous uncultivable land and hits the poor farmers hardest, ultimately unemployment and under employment are the causes of direct outcome of degradation. Finally the farmers were forced to sale their agricultural based assets.

The table 5.22 explains that the farmers who had involved in sale of their assets. It revealed that 29 percent of the farmers sold their part of land. Of these the Zone-I (24.00%) was higher than zone-II (5.33 percent). The reasons can be traced that as the zone-I come closer to the city Tiruppur, their land in those places required for industrial purposes, hence the farmers waiting for their turn to sale the land, as the land not supplementing any income. But this is not the case of zone-II as it was away from the cities. Similarly 25 percent of the farmers sold their livestock as their cattle were receiving disease and for supplementing their income level. Nearly 16 percent of the farmers, of which 22 percent from Zone-II had involved in sale of agricultural tools and machineries, like tractor, pump sets etc. Similarly 17 percent of the farmers had involved sale of other small assets as their income level was on declining.

Hypothesis testing

- There is no association between the effluent water irrigation and sale of agricultural based assets.

To test the above hypothesis, a Non-Parametric test viz., Chi-Square is applied. The study of chi-square is estimated to find out the degree of discrepancy between observed frequencies and theoretical frequencies and thus to determine whether the discrepancy so obtained between observed frequencies and theoretical frequencies in due to error of sampling or due to chance.

$$\Psi^2 = \Sigma \frac{(O - E)^2}{E}$$

O = Observed Frequency of the event

E = Expected Frequency of the Event

The result of the study is as follows,

Estimated Chi-square value – 9.974

Degrees of Freedom - 3

Table value - 7.815

Since the calculated value of Ψ^2 = 9.974 is higher than the table value Ψ^2 for 3 degrees of freedom=7.185 the null hypothesis is rejected at 5% level i.e due to the effluent water irrigation, the activities involvement are associated with the sale of assets.

Table 5.23

Distribution of Respondents by Distress Sales in Marketing

Sl. No.	*Zone*	*Cereals*		*Coconut*		*Cotton*		*Vegetables*		*Others*	
		Yes	*No*	*Yes*	*No*	*Yes*	*No*	*Yes*	*No*	*Yes*	*No*
1	Zone-I	58 (77.3)	Nil	2 (2.66)	Nil	7 (9.33)	Nil	6 (8.00)	Nil	21 (2.66)	Nil
2	Zone-II	63 (84)	Nil	5 (6.66)	Nil	3 (4.06)	Nil	2 (12.36)	Nil	2 (2.66)	Nil
	Total	121 (80.65)	0%	7 (4.60)	0%	10 (6.67)	0%	8 (5.33)	0%	4% (2.60)	0%

Figure in () Percentages.

Marketing is the performance of all business activities involved in the flow of goods and services form the point of initial of agricultural production until they are in the hands of the ultimate consumer. In these areas, that the agricultural production is below subsistence level and the marketing activities is rather unseen.

Production techniques, input output operations, harvesting, transporting, marketing and integration should be done on a scientific basis. The operational unit will be kept economics and the entire agricultural produce is considered as "produce for marketing" like the industries. But in the study area only a few large scale farmers were able to produce some what quantities of surplus production, while not all the producible goods by small and marginal farmers, comes to the market. Out of the total production, the farmers retain major portion of the produce for their own consumption, fodder for livestock and set apart as seed for the next sowing season. Many small farmers reported that they were not able to cultivate land as commercial motives, but only they retain the image of the farmers and as such the produces from the land not to meet their family consumption and they were forced to buy the seeds with high cost for the survival of their livestock.

The table 5.23 analyses the farmers, who were all had distress sale, in marketing in different crop cultivations. It showed and disheartened that almost 100 percent of the respondent not positively inclined about their sales of their produce. Of them, 81 percent of total millet producers were negative sales as the maize crop is not cultivated for price gains, but for family consumption and as a fodder for livestock.

Table 5.24

Opinion of Respondents on Roll of Panchayat Union

Sl.No.	*Zone*	*Zone-I*	*Zone-II*	*% out of Total*
1	Effective	32(42.66)	34(45.33)	9(12.00)
2	Insufficient	45(60.00)	22(29.33)	8(10.66)
3	Poor	77(51.33)	56(37.33)	17(11.00)

N = 150 Figure in () Percentages.

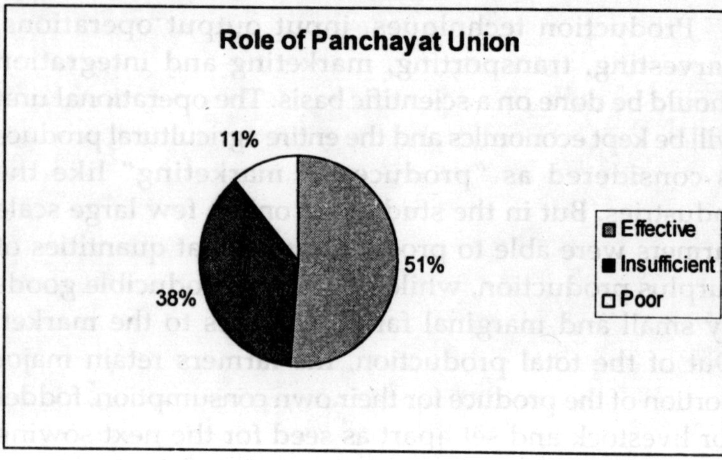

In each reference village in study area there were number of hamlets affected by the river Noyyal and they had to receive, bring (or) buy drinking water from other places. Pollution and scarce water supply in the regions created a critical and complex water situation. The respective panchayat have been taking concrete steps, but they could not produce positive results. In the opinion of the farmers nearly 51 percent had reported that the role of Panchayat for taking steps were adequate, while 37 percent of them opined that they were insufficient and 11 percent had a negative opinion of the role of their Panchayat.

Table 5.25

Opinion of Farmers Association in Study Area

Sl.No.	*Criteria of Opinion*	*Zone – I*	*Zone - II*	*% ou' of Total*
1	Highly Commendable	51(68.00)	58(77.33)	109(72.66)
2	Commendable	22(29.33)	16(21.33)	38(25.33)
3	Poor	2(2.66)	1(1.33)	3(2.00)

Figure in () Percentages.

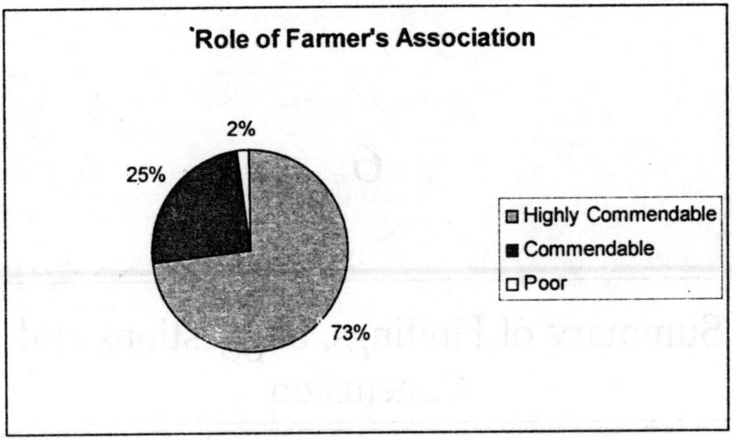

There are number of farmers association are functioning to represent the problem of farmers. Almost all the farmers were a membership in these associations. For many years the villager farmers have been protesting against the pollution of the river. The farmers association helped the farmers and villagers bring the struggle on to the negotiations table with the government. On behalf of the farmers, the association has filed a petition on High court, Chennai for compensation. It could be identified that from the table5.25the farmers association were doing an effective role, and they play a highly commendable rolc (72.66%) in those areas. Where as 25 percent of the farmers had a mind on the farmer association was doing a commendable work. Only 2 percent of them had a negative opinion about the farmers association.

6

Summary of Findings, Suggestions and Conclusion

This chapter summarizes the study as a whole. The major findings of the analysis of impact of effluent water irrigation are presented in the section. Further few suggestions are offered for the socio- economic development of the farmers and devising strategies to the policy makers and the textile units.

Major Findings

a. Demographic Profiles of the farmer

- Among the respondents taken to the study, majority of respondents (47.34%) had the family member of 5 to 8.
- It was identified that majority of the farmers, who have been samples, were residing in the study area (87%) for more than 30 years.
- The study had found out that majority of the (52%) farmers were having less than 3 acres of land they are all otherwise know as small scale and marginal farmer. More over 90 percent of

the farmer had been cultivating land for more than 20 years.

b. Impact of Effluent Water on Agricultural Land

- The percolation of effluent water affects the quality of land, and there by the source of livelihood of the people. It could be identified that only 10 percent of the total farmers, taken to study, depending agricultural land as a major source of livelihood, while before the impact 72 percent of them had been depending the agricultural activities. The farmers were supplemented their basic daily requirement from non-farm economic activities..
- With regard to the sources of water, none of the farmers depend tank water for irrigation as the Orathupalayam dam became a polluted water tank. But 51percent of them depended earlier. However the farmer, who depend rain water, had increased.
- In respect of impact of effluent, 65 percent of farmers in zone I, and 80 percent from zone II were affected extremely by the contaminated water. Of them, the small and marginal farmers had to pay the heavy price for irrigating the effluent water.
- Yielding per hectare of lands was drastically reduced. The crops cultivators such as millet, coconut and cotton acknowledged that their yielding from land were low. The soils in these areas were highly exhausted, because of the contaminated water.

- The data pertaining to the years of flow of water revealed that 79 percent of the farmers in zone I and 83 percent of the farmers in zone II were receiving the problem for more than 15 years and 8 to 15 years respectively.
- With regard to sources of percolation of waste water 62 percent of the farmers land were polluted by direct flow of effluent water from Noyyal River, where as 47 percent of the land were contaminated by Orathupalayam reservoir.
- The condition of land by the small and marginal farmers was mostly very poor. Of the total farmers 83 percent in zone II were highly polluted. Invariably all the farmers reported that their land had lost its nutrition that was the principle cause of downsizing the productivity of land.
- It was understood that majority (80%) of all the categories of farmers had indicated that they shifted from intensive crop cultivation to rain-fed crop cultivation.

c. *Present Status of Farmers in Affected Region*

- The farmers were seemingly engaged in agricultural activities and contribute little to agricultural production. It was understood that the farmers were affected by yielding less of few crops, reduction of amount by doing agricultural activities.
- The season of cultivation had a drastic change. An average of 65 percent of respondents was able to cultivate the land at 3 seasons per year before the pollution impact. At present it was

reduced to only 9 percent and most of the farmers have been cultivating the land for only one season.

- As the effluent water irrigation is a persistent problem, 29 percent of the farmers sold portion of their land and most of the farmers in Tiruppur area were ready to sell, as the agricultural land in those places are required for industrial establishments.
- Out of the total production, the farmer retain major portion of their produce for their own consumption and fodder for livestock. In many cases, the small and marginal farmers were forced to buy the seeds with high cost for the survival of live stocks.
- The study had found that almost all the farmers had a distress sale in marketing, and in many cases most of the small and marginal farmers did not bring their product to market for price gains.
- All the farmers were members in farmers' movements, and 73 percent had an opinion that the activities taken up for solving the problem by the association were highly commendable.
- Before the impact of effluent water, the crops grown in the study area were cotton, herbal (turmeric), onion, and coconuts. At present the maize was the only major crop cultivated. Rice and vegetable fields almost disappeared and cotton and coconut were in the verge of disappearance.

Suggestions

In view of the serious socio-economic repercussions of the effluent water, they should not be allowed to continue for further degrading the ecology of the three districts. They should be prevented and immediate action is needed from all the responsible functionaries. As the impact levels vary from one to another area, the remedies should be undertaken with reference to the particular causes. The following suggestions can help the farmers and the stakeholders as well.

a. Recycle of Effluent Water

1. Given the environmental problem caused by effluent irrigation, recycling the waste water in industrial sector could be the option. Since high TDS concentration is the major problem of the textile units, these units need to decrease TDS by Reverse Osmosis (RO) or other technologies. Industries can also reduce the pollutions load in their effluents through cleaner production technologies which consume less water and chemicals.
2. All the Common Effluent Treatment Plants shall necessary install reverse osmosis / nano system and make adequate arrangement to transfer the permeate back to the member units. The evaporate system shall also directly receive RO rejects from the individual units on charge basis. The economic feasibility about the total number and sizes of the units shall have to be worked out at the time of preparing detailed project report.
3. Those units who are neither discharging then

effluents with the CETPs nor having their own R.O/ nano systems in place shall have to either close down (or) shift operations with in the reach of CETPs.

4. There are 8 CETP established in Trippur area, of which only 2 have been functioning at present. According to the study of Loss of Ecology and Authority (LEA) the total loss due to the effluent water exceeds Rs 100 crores. Further damage to the environment and ecology can be arrested only by upgrading the CETPs.
5. Approximately an estimated amount of Rs 200 crores has been spent for this purchase of fresh water through lorries. If water is recycled through modernization or upgradation through reserve osmosis the amount can be saved. This saving can be effectively utilized for industrial development purpose to meet the potential competitors in the emerging open market.
6. Annual agricultural productivity loss in villages affected by the industrial activities in Tiruppur only amounts to around 20 crores. By establishing CETP, both damages of the environment and economic loss can be averted. In addition to this the reclamation of orathupalayam dam and Noyyal River is made feasible by CETPs. So that agriculture Production can be increased.
7. By establishing CETPs it may be possible for integrating the environmental concern in economic and social development. This can not only takes care of the environmental pollution but also increase the price competitiveness

through effective utilization of natural resources and take care of economic development of Tiruppur.

8. It is imperative that the central and state government to take immediate steps for necessary intervention by implementing zero discharge system in order to avoid further degradation of land. By implementing the zero discharge system there are multiple economic and environmental benefits will be to make both the industries as well as the rural areas. They are listed below:
   - Ground water pollution as well as ground water depletion in the surrounding areas is greatly minimized.
   - Production of hazardous waste can be averted by the elimination of conventional treatment system and upgradation of the zero discharge system with salt recovery and reuse.
   - Reuse of water and salt through the implementation of the zero discharge system would greatly reduce the processing expenses. Thus the dyeing and bleaching units in Tiruppur can become potentially competitors.

B. Stakeholders Effort

1. It is imperative for the government to make policy decision which might have long standing industrial, social and environmental impacts. The government at first should decide what can be done to control industrial pollution by

appropriately identifying the necessary waste water treatment scheme, study the technical feasibility of the project, define the goals and time frame for the implementation of the project, and should allocate necessary funding for achieving the defined objectives.

2. Recharge of fresh water through to traditional as well as modern rain water harvesting method will help to reduce the level of pollution through dilution. In this respect construction of more check dams and percolation ponds and reclamation of tanks and other degraded water sources can help to overcome this problem
3. Creating more awareness regarding the adverse consequences of effluent irrigation among different stake holders viz, industrialists, farmers, NGOs, and members from PCB. Collective efforts towards pollution management have not taken place in many parts of the area. The local NGOs must play a constructive role in addressing the problem and to find a solution to the problems faced by the farmers.
4. In the areas around the Orathupalayam reservoir, the water quality and ground water was severely deteriorated, some scheme towards remediation shall be planned and executed for a fast recovery of ground water quality. The scheme should also explore the feasibility of ground water recharging through low TDS water. The entire on going programme towards cleaning of Orathupalayam reservoir bed at directed by hon'ble High court and other strategies adopted by government of Tamilnadu in their regards shall continue with

time bound target towards execution of schemes.

5. The agriculture department of the state government should take necessary steps to supply the seeds and other agriculture inputs like fertilizers and pesticides at a subsidised rate so that the farmers can ensure better agriculture production.
6. The farmers as well as other villagers and their live stock met serious health problems due to the drinking of the contaminated water. It is essential to conduct health care camps once in a month in these affected regions. The farmers association should play a constructive role in addressing the problem of the farmers.
7. According to many farmers, they did not receive the compensation amount, which was ordered by the Hon'ble High court and the amount as recommended by the Anna University, till now the farmers have received only partial amount. In this regards the stakeholders must take the necessary steps for receiving the amount.
8. Invariably, the farmers in the affected region did not bring their products to market. The government can announce a reasonable support price for procuring their agriculture products.

Conclusion

The use of the untreated effluent water from the Noyyal River for the irrigation of crops created adverse impact on the farming lands, as well as it increases pressure on environment by damaging the ecology. The study had found that none of the rural people were now dependent upon on the river water for drinking, irrigation, fishing, washing the cloths, pebble collection, bathing the livestock. Besides it has polluted all the other drinking water sources. The rural people had to go few Kilometers for fetching their drinking water. Agricultural works is closely disappearing, and if remedial action and the preventing steps will not be taken in due course, soon disappear. Although the people were aware of the water pollution their awareness was not yet translated into action which was mainly because of lack of integrated institutional mechanism. A wide public participation with effective institution support could help in controlling the discharge of industrial effluent into the river, towards abatement and control of river pollution and make the agricultural sector and the dependent more progress

Bibliography

Books

1. Virmani, S.M., Katyal, Eswaran H, & Abrol, I.P., *Stressed Ecosystem and Sustainable Agriculture*, Oxford and IBH Publications, New Delhi.
2. Desai, R.G. (2001) *Agricultural Economics*, Himalaya Publishing House, New Delhi.
3. Jagadees Pandi, S. (2008) *Incidence of Sickness in Small Scale Units*, Discovery Publishers, New Delhi.
4. Karpagam, M. (1991) *Environmental Economics*, Sterling Publishers, New Delhi.
5. Kudesia (2003) *Water Pollution*, Pragat Prakkash Publishers, New Delhi,
6. Manjunath, D.L. (2007) *Environmental Studies*, Dorling Kindersley, New delhi.
7. Rana, S.V.S. (2006) *Environmental Pollution- Health and Toxicology*, Narosa Publishing House, New Delhi.
8. Senaca, J.J. & Taussig, M.K. (1984) *Environmental Economics*, Prentice-Hall Inc, New Jersey.
9. Sehgal, J. & Abrol, I.P. (1994) *Soil Degradation in India—Status and Impact*, Oxford and IBH Publishing Co.
10. Vasudevan, N. (2006) *Essentials of Environmental Science*, Narosa Publishing House, New Delhi.

Journals & Magazines

1. Ahamed A. Alam (2003) "Physio-chemical and toxicological study of Industrial Effluent in and around Delhi", *Chemical Environment Research,* Vol. 12, No. 1 & 2, pp. 5-13.

11. Ali Hariwandi (1982) "The use of effluent water for Turf Craft irrigation", *Cooperative Extension,* vol. 32, No. 3 & 4.

12. Angelora,V., Janov, R., Delibaltora, V. (2004) "Bio-accumulation and distribution of heavy metals in fibre crops", *Industrial Crops and Products,* Vol. 19, pp. 197-205.

13. Aparna S. & Anandavalli (1984) "A study on HO_2 pollution by tanneries in Dindigul", *Indian Jounal of Pollution Control,* Vol., No. 2.

14. Bhagirath Behara, Ratna Reddy V. (2002), "Environment and Accountability – Impact of Industrial pollution on Rural Communities", *Economic and Political Weekly,* Vol. 37, No. 3, January.

15. Chellappan Sekar (2003) "Effect of Industrial Effluents on ground water and the consequent impact on Economic, Social, Environmental, Governance", Paper presented a National Seminar on Water, Centre for Economic and Social Studies, Hyderabad, July 30-31.

16. Gaurav Dwivedi "Tiruppur water supply and Sewage project", *News for Sustainable Development, News,* No. 17, Bangalore.

17. Fazeli, M.S., Khesravan, M., Hosini, S. Satyanarayan, P.N. Sathis, "Enrichments of heavy metals in Paddy crops irrigated by paper mill effluent near Nanjangud Mysore Dt"., *Environmentul Geology,* Vol. 34, No. 4, June 1998.

18. Food and Agricultural Organisation –FAO(1976) "A Framework for Land Evaluation", *FAO Soils Bulletin,* Rome pp. 32.

19. Halliwal D.J., Barlow, K.M. & Nash, D.M. (2002) "A review of the effects of waste water sodium in soil properties and their implications for irrigation system", *Australian Journal of Soil Research,* vol. 39, pp. 1259-1267.

20. Jain, A.V., Gupta, I.C. (1993) "Impact of Textile Industrial effluents on Agricultural Land- A Case Study", *Indian Journal of Environmental & Health,* Vol. 30, No. 2, pp. 132-138.

21. Joshi and Jha (1991) "An Economic Enquiry into the Impact

of Soil alkalinity and Water logging" *Indian Journal of Agricultural Economics,* Vol. 47, pp. 195-204.

22. Jothimani, R., Baskaran, B. (2002) "Effluents of dilution and dynamics of physical factor during factory Effluent irrigation", *Environmental Monitoring,* Vol. 12, No. 4, pp. 225.
23. Kunvar P. Singh, Dinesh Mohan, Sarita Sinha, Dalwani, (2004) "Impact assessment of treated/ untreated waste water toxicants discharged by sewage treatment plants on Health, Agricultural and environmental quality in the waste water disposal area", *Chemosphere,* Vol. 55, issue 2, April, pp. 227-255.
24. Malik S. Negi (2003) "Impact of distillery effluent in water quality and phytoplankton in river saving", *Indian Journal of Environment and Science,* Vol. 7, No. 2, pp. 163-167.
25. Nagaraj, N., Chandrasekar, H., Mangala K.P. (2003) "Groundwater Pollution induced Externalities in Vrishabavathi River basin in Bangalore Hunter Lands : A Study in Peninsular India", Paper Presented at Information support sustainable water Management: From Local to Global Level, Netherland 15-18 September.
26. Ninan, K.N (2002) "Watershed Development Programmes in India – A Review" Paper presented at the ISCO conference, Beijing.
27. Ofuso, N., Asiedu, K., Otieno, N.A., Omolo, Ethengi (1999) "Sewage reuse for irrigation in Athi river Town Kenya : Its implications on public health", *Water Science Technology,* Vol. 33, pp. 343-346.
28. Patel, K.P. and Pandya, R.P. (2003) "Sustainability of Industrial Effluents for irrigation around Bharuch and Angalshwar Industrial Zone of Gujarat", *Pollution Research,* Vol. 22, No. 22, pp. 241-245.
29. Phiri, O., Mumba, P., Moyo, Z., Kaderva, W. (2005) "Assessment of the Industrial effluents on water quality of receiving river in Urban areas in Malawi", *Science and Technology,* Vol. 2, No. 33, pp. 237-244.
30. Rao, A.V, Jain, A.V., Gupta, I.D., (1993) "Impact of Textile Industrial effluents on agricultural Land-A case study", *Indian Journal of Environmental Health,* Vol. 30, No. 2, pp. 132-138.
31. Ratna Reddy, V. & Bhagirath Behera (2006) "Impact of Water pollution on rural communication--An Economic Analysis",

Ecology Economics, Vol. 58, Issue 3, June.

32. Satyamoorthy (2006) "Tiruppur Effluent discharge – PAC marks 800 on scheme", *The Hindu* (Daily News) November 2, Chennai.
33. Srinivas Mudrakastra, Jain Sheth and Srinath, J. (2006) "Unclogging the Khari River in Ahmedabad – Stakeholders come together to halt pollution", *Economic and Political Weekly*, Vol. 41, No. 7, February-18, pp. 587.
34. Sridhar (2005) "The farmers plight" *Front Line* (News Magazine), vol. 22 No. 17, August.
35. The Editor (2001) "Industries polluting various Rivers in Tamilnadu", *The Times of India* (News Daily) November, New Delhi.
36. The Editor (2006) "High Court Monitory Committee visits Orathupalayam Dam", *The Hindu* (News Daily) August 18 Erode.
37. Thomas Jacob, Jayabal, Azariah (1997) "Environmental perception of Textile Industrial Pollution, in Tiruppur India" *Eubios Journal of Asian and International Bio-ethics*, vol. 7, pp. 162-165.
38. The Editor (2004) " Tiruppur dye sector in pollution – told to Folk out Rs. 104, or compensation", *The Hindu* (News Daily) March 17, Chennai.
39. Udaya Suriya and Prabhu, P.C. (2003) "Impact of treated paper mill effluent irrigation on quality of soil and ground water", *Madras Agricultural Journal*, Vol. 90, No. 7, pp. 540-541.
40. Yadav, B.K. Lourdraj and Paul Sebastian (1998) "A Review on Industrial effluents use in agriculture– Problems and Prospects", *Crop Research*, Vol. 31 (2), pp. 183-191.

Reports/Working Papers

1. Azez P.A. (2001) " Environmental implication of untreated effluent from bleaching and Dyeing, ed. Senthilnathan, Eco friendly technology for waste minimization in Textile Industry," Centre for Environment Education, New Delhi.
2. Balaji S. (2003) State Environment Report of Tamilnadu, Chennai.
3. Banarjie and Munshi, K. (2003) "How efficiency in capital

allocated? Evidence from the knitted garment Industry in Tiruppur", Bureau for Research in Economics Analysis and Development, working paper No. 004 Massachusetts Institute of Technology.

4. Bhamoria (2004) "Waste water irrigation in Vadodara, Gujarat; Economic Catalyst for marginalized communities" (ed), Christopher Scott, Fariqui and Liqa Rashid, *Waster Water use in irrigated Agriculture- Confirming the livelihood and Environment Realties,* The International Development Research Centre, Ottowa.

5. Crow, M.T. (1999) "Successfully Adjusting the Environmental Regulation; The small firm cluster of Tiruppur; India", Master in City planning, Massachusetts Institute in Technology pp118.

6. Eswaramoorthi & Dhanabal (2005) "Textile effluent and waste water management in Tiruppur India, need for Government Aid" Report prepared by Environment with People's Involvement& Co-ordination in India,(NGO), August.

7. Food and Agricultural Organisation of the United Nations (1994) "Land Degradation in South Asia: Its Severity, causes and effects upon people, World Soil Resources Report, Rome.

8. Govindarajulu (2003) "Industrial Effluent and Health status--A case study of Noyyal River Basin" Proceeding of the Third International Conference on Environment and Health, Chennai, December 15-17, pp. 150-157.

9. Government of India (1995), Ministry of Finance, Economic Survey Report.

10. Government of India (2006) "Compendium of Environment Statistics in India", Ministry of Statistics and Implementation Programme, pp. 1-3.

11. Government of India (2005) Tamilnadu Development Report. Planning Commission in India, New Delhi.

12. Government of Tamilnadu (2006) "Environmental Concern of Tamilnadu/Agriculture" State Environment Report. Department of Environment, Chennai.

13. Government of Tamilnadu (2007) "Unpublished record of Tamilnadu Pollution Control Board", Tiruppur Branch.

14. Profile of Coimbatore District (2005) District Collectorate

Office, Coimbatore

15. Maria A. (2003) "The cost of water pollution in India", Paper presented at Conference on market development on waste and waste technologies through environmental economics, New Delhi, 30-31 October.

16. Palanivel & Rajagopalan (1999) "The present status of the Noyyal river" Proceedings of the workshop on Environmental Status of River in Tamil Nadu", Bharathiar University, Coimbatore.

17. Paul Appasamy & Prakash Nelliyat (2007) "Compensating the loss of eco system services due to pollution in Noyyal River basin " Working paper 14/2006, Madras School of Economics, Chennai.

18. Prakash Nelliyat (2005) Industrial growth and water pollution in Noyyal River Basis, Working Paper Series, Madras School of Economics Chennai.

19. Sankar Raj, L. Subramanyam T.P., Siddhamalaii A & Faroque Ahamed N. (2002) "Quality of Soil and Water for agriculture in Noyyal River basin. Tamilnadu" Soil Survey and Land use Organization, Department of Agriculture, Coimbatore.

20. Sakthivel, A. (2004) "Tiruppur Knitwear Export Cluster, India ITC Executives Forum on "National Export Strategies – competitiveness through public private partnership success and lessons learned", Montrex, Switzerland September 26-29.

21. Sachidhanantha Mukerjee & Prakash Nelliyat (2007) "ground water pollution and emerging environmental challenges of Industrial effluents irrigation 'A case study in Mettupalayam Taluk, T.N. "Paper presented at IWMI – TATA water policy programme, 5th Annual Research meet". Institute of Rural Management, Anand, Gujarat.

22. Senthil Nathan, S. (1998) "Environmental impacts of dyeing and bleaching and Dyeing Industrial Effluents on abiotic and biotic components on river Noyyal", Paper presented at the Seminar on Impact of Industrial effluents, Sewage and agricultural practice on water quality, by Public Works Department, Coimbatore, August.

23. Simon Toze (2004): "Reuse of effluent water benefits and Risks", Paper presented at International Crop Science Congress, September, Brisbane, Australia.

Websites

1. Wikipedia of Enclocycopedia available at www.wikepedia.org
2. Krishna Furn (2004) Effects of dyeing and bleaching industries on the area around the Orathupalayam Dam, Southern India, http://www.w-program.nu/
3. Government of India, Ministry of Environment and Forest available at www.enfor.nic.in
4. Centre for Science and Environment available at www.rainwaterharvesting.org
5. Sacred land film project –. http://www.sacredland.org/world_sites_pages/Ganges.html

❑❑❑

Index

◆◆◆